CLONES and CLONES

also by Martha C. Nussbaum

The Fragility of Goodness
Love's Knowledge
Cultivating Humanity

also by Cass R. Sunstein

Free Markets and Social Justice
Legal Reasoning and Political Conflict
Democracy and the Problem of Free Speech

CLONES and CLONES

Facts and Fantasies About Human Cloning

edited by

Martha C. Nussbaum

and

Cass R. Sunstein

W · W · Norton & Company
New York   London

For information about permission to reproduce selections from this book, write to Permissions, W. W. Norton & Company, Inc. 500 Fifth Avenue, New York, NY 10110.

The text of this book is composed in 12/14 Perpetua
with the display set in Bodega Serif Medium and Light
Composition and manufacturing by The Haddon Craftsmen, Inc.
Book design by Margaret M. Wagner

Library of Congress Cataloging-in-Publication Data
Clones and clones : facts and fantasies about human cloning / edited
by Martha C. Nussbaum and Cass R. Sunstein.
p. cm.
Includes bibliographical references.
ISBN 0-393-04648-6
1. Cloning—Social aspects. 2. Cloning—Moral and ethical
aspects. 3. Human reproductive technology—Social aspects.
4. Human reproductive technology—Moral and ethical aspects.
5. Human genetics—Social aspects. 6. Human genetics—Moral and
ethical aspects. I. Nussbaum, Martha Craven, 1947–   .
II. Sunstein, Cass R.
QH442.2.C55   1998
174'.25—dc21   97-51781
                                                              CIP

ISBN 0-393-32001-4 pbk.

W. W. Norton & Company, Inc., 500 Fifth Avenue, New York, N.Y. 10110
http://www.wwnorton.com

W. W. Norton & Company Ltd., 10 Coptic Street, London WC1A 1PU

1 2 3 4 5 6 7 8 9 0

In memory of
Marian Goodrich Sunstein,
1917–1997

Contents

PART III   ETHICS AND RELIGION

PART IV   LAW AND PUBLIC POLICY

PART V   FICTION AND FANTASY

Introduction

When Ian Wilmut and his colleagues at the Roslin Institute announced the successful cloning of a sheep from the mammary cells of an adult female, the world reacted with intense emotion. Experiments in cloning had been going on for at least forty years. But the arrival of Dolly made it clear that human beings would soon have to face the possibility of human cloning—and it has been this idea, far more than the reality of animal cloning, that has caused public anxiety. To many if not most of us, cloning represents a possible turning point in the history of humanity. Some view the prospect with alarm; some with disgust; some with joy; some with grief for the life we used to have, and will shortly have no longer. Some, too, are calm and matter-of-fact about the entire affair, urging us to let science take its course before we conclude that dreadful things are at hand. But almost everyone is asking questions.

What is all the emotion about? Human beings have always been afraid of their own creative power, and the idea of man-made life was scary long before science was in any position to think of procedures that might make it reality. The Greek god Hephaistos, god of artifice and metalworking, cleverly stocked his workshop with robots, making anthropomorphic metal creatures to be his slaves and lighten his burden. He fashioned tripods that could move of their own accord, rolling off to the assembly of the gods and back again, "a wonder to behold." Stranger still, he also created young women out of gold, "just like living girls," with "intellect" and "voice" and even "strength," to help him in his labors.[1] These agile workers were human bodies without regular human souls, children without parents, neither human nor bestial nor even divine. To the Greeks already, then, the idea that there was a "natural" way of reproducing, and that creatures who bypassed that route were a little peculiar, came quite easily, without science to back it up. Not that the gods and heroes were without their own peculiarities, choosing such strange means of birth as being released from the forehead with a hammer (Athena), being sewn up in one's father's thigh (Dionysus),[2] getting children out of some decayed old dragon teeth that happened to fall into the fertile earth of Thebes (the Theban "sown men"),[3] and even, with a simple economy of means, masturbating into the fertile earth of Athens to create a nation of male clones without bothersome mothers. (The Athenians were thus, in their own civic mythology, a nation descended from male clones, very proud to have no jot of the female in their makeup.)[4] The gods, in short, were clearly not convinced that nature had ordained just one way for human beings to be born, and they tirelessly experimented with new technologies. But gods and heroes are free from many constraints that they impose on mere mortals. Prometheus got his liver eaten out by an eagle for giving us fire, thence metalworking and other creative arts; and the idea that new arts are all new transgressions, each bringing divine punishment in its train, is a very old idea in Western thought about what it is to be human.

The scientific advances of modernity, however, gave these anxieties new intensity and specificity. The idea of a life-form made in a laboratory that would ultimately destroy its creator underlies the horror of *Frankenstein,* a tale made profound by the monster's noble simplicity and its victimization by the "normal" among us. The idea that we might make a human being without a soul was a source of horror; equally horrible, however, was the thought that we might make a being who *did* have a soul, but who, by virtue of its strange origin, would be doomed to live without love. More recently, the immensely popular film *Invasion of the Body Snatchers* associated abnormal man-made life with the soulless automata that Americans saw when they imagined a life under Communism. The people who came out of the pods were, in effect, clones of the people whose lives they took over. They looked like us, and kind of talked like us. But they didn't like jazz, they never got angry or frightened, and when you kissed one of them, then you really knew what fear was. Sex, freedom, anger, being American, and music were thus lined up on one side; abnormal origins, totalitarianism, soulless passivity, being foreign, and nonmusic were carefully arrayed on the other. What's to choose? We have to stop them, before they stop us.

These are some of the nightmares and fantasies that underlie the intense outpouring of negative emotion in response to Dolly's life. We want to know what human cloning would mean for the children born in this way. Would they really be creatures without souls, not fully human? Even if they weren't, would we treat them this way? What would cloning do to the parent-child relationship? What kind of lives could, or would, clones have? If we could choose the genetic makeup of a child, would unconditional love for children become rarer than it is now? And what would the option of cloning do to us more generally? Would we stop wanting to have sex? Want sex only for superficial self-gratification? Or would we go on wanting to express love and friendship through sex? Would people's sense of partnership even lead them to prefer natural children to clones, since they would be the mingling of the genetic equipment of both parents? What is the relationship between cloning and our various religious traditions?

Who would choose cloning—the infertile, the narcissistic, people with a loss they want to make good, people with a grudge against humanity, people who hate chance? Whom would they choose to clone? Themselves? Their favorite basketball star? Hitler? Gandhi? Mozart? And what would become of our world, with dozens of Hitlers running around, opposed by hundreds of Gandhis? Could we make it better? Wouldn't it almost surely be made worse? And what about basketball and music—surely they wouldn't remain the same either. It's comforting to think of dozens of Mozarts and Beethovens, since music seems to be one of those goods that expands without limit, and admits of no diminution through excess. But what about a National Basketball Association filled with teams of Michael Jordans (with a Scottie Pippen and a Dennis Rodman on each, we may hope, or fear, and a Phil Jackson coaching everyone)? Would that still be the same game? Or wouldn't the limits of the human body against which talented athletes strive lose their meaning when we could just make another Michael Jordan any time we wanted to?

These questions can be posed by science, and science can give us the real facts, telling us that many of the possibilities we envisage are very unrealistic. We attempt to outline the basic facts about cloning here, and in a way that will be easily intelligible to nonspecialists. Thus, the distinguished scientists in this volume remind us that clones occur already in nature: identical twins. They also remind us that nature's form of cloning hasn't yet produced horror movie scenarios, except in horror movies. Science can also shed a great deal of light on a much-discussed issue, the interaction of genetic endowment with environment, and scientists may remind us—as do several writers in our first section here—that a clone is probably going to end up very far indeed from being the same person, even farther than are identical twins with separate upbringings, given that clones will also be born into different generations. In these many ways science can set us straight about what questions we should really be asking.

Science, however, doesn't give us the answers to the ethical, political, social, and religious questions raised by cloning. These answers need to be worked out, ultimately, in the course of public debate. But the humanities and the social sciences can help us lay out the options in a clear way, and give us some good arguments to ponder. Our second section brings together a group of commentators from a variety of social perspectives. They think about the relationship of cloning to the psychological development of children; to our most basic feelings about our bodies and their products; to feminism and to homosexuality; to myths of the dangerous or useful double.

Our third section turns to normative argument in ethics and religion. The authors analyze the major ethical arguments that bear on the decision whether human cloning should be permitted and ask what insights several major religious traditions offer us about the strange future we may face.

In our fourth section we turn to issues of law and public policy, as specialists in economics, sociology, and law think about whether governments should allow human beings to be cloned. They also discuss the implications of cloning, and any bans on cloning, for children and actual or prospective parents; for the freedom of scientific inquiry; for constitutional debates about privacy and equality; for the position of existing citizens who challenge current social attitudes about sexuality; for the fertility of the species; and for other ethical issues concerning the family. Some of the arguments of this and the previous section played a role in the controversial report of the National Bioethics Advisory Commission, whose recommendations we reprint here.

But cloning, so far, is still in our future. Like the Greeks, therefore, we also feel the need for fiction and fantasy, to map out some alternative futures for us with the imagination's flexibility and precision. We therefore end with a poem and three short stories—one by a writer of science fiction and two by fiction-writing philosophers. Deliberative, elegiac, horrifying, and happy, these pieces map the trajectory of our sentiments about cloning, even as they ponder its possibilities.

M.C.N.

C.R.S.

Chicago, October 1997

Endnotes

1. For the tripods, see *Iliad* 18.375; for the young women, 18.417–20.

2. For the double birth of Dionysus (once from his mother, Semele, once from Zeus's thigh), see Eruipides, *Bacchae,* 88–98. Zeus pinned the young infant up in his thigh, creating an artificial womb, to hide him from Hera, who was angry at his adultery.

3. This story of the "sown men" is central to the plot of Aeschylus's *Seven Against Thebes,* and is retold in Plato's *Republic* as a crucial part of Socrates's "noble lie" about the origins of the citizens of the ideal city.

4. On myths of misogyny connected with the idea that Athenian males were created without sexual reproduction as a result of Erechtheus's masturbation, see Nicole Loraux, *The Children of Athena: Athenian Ideas About Citizenship and the Division Between the Sexes* (Princeton: Princeton University Press, 1993; originally published in French, Paris 1981).

Acknowledgments

We are very grateful to our secretaries, Shirley Evans and Marlene Vellinga, for excellent work in organizing a somewhat unruly editorial process. We are also grateful to our outstanding editor, Alane Mason, who provided overall guidance and wisdom and who offered superb suggestions throughout.

Cass Sunstein's mother, Marian Goodrich Sunstein, became ill and died as this book was nearing completion. She knew about the project and found it a particular source of fascination and delight; perhaps her son may be forgiven for saying that it was a great gift to be able to discuss it with her in her final months.

M.C.N.
C.R.S
December 1, 1997

PART I
Science

Viable Offspring Derived from Fetal and Adult Mammalian Cells

I. Wilmut, A. E. Schnieke, J. McWhir, A. J. Kind, and K. H. S. Campbell

ertilization of mammalian eggs is followed by successive cell divisions and progressive differentiation, first into the early embryo and subsequently into all of the cell types that make up the adult animal. Transfer of a single nucleus at a specific stage of development, to an enucleated unfertilized egg, provided an opportunity to investigate whether cellular differentiation to that stage involved irreversible genetic modification. The first offspring to develop from a differentiated cell were born after nuclear transfer from an embryo-derived cell line that had been induced to become quiescent.[1] Using the same procedure, we now report the birth of live lambs from three new cell populations established from adult mammary gland, fetus and embryo. The fact that a lamb was derived from an adult cell confirms that differentiation of that cell did not involve the irreversible modification of genetic material required for development to term. The birth of lambs from differentiated fetal and adult cells also reinforces previous speculation[1,2] that by inducing donor cells to become quiescent it will be possible to obtain normal development from a wide variety of differentiated cells.

It has long been known that in amphibians, nuclei transferred from adult keratinocytes established in culture support development to the juvenile, tadpole stage.[3] Although this involves differentiation into complex tissues and organs, no development to the adult stage was reported, leaving open the question of whether a differnetiated adult nucleus can be fully reprogrammed. Previously we reported the birth of live lambs after nuclear transfer from cultured embryonic cells that had been induced into quiescence. We suggested that inducing the donor cell to exit the growth phase causes changes in chromatin structure that facilitate reprogramming of gene expression and that development would be normal if nuclei are used from a variety of differentiated donor cells in similar regimes. Here we investigate whether normal development to term is possible when donor cells derived from fetal or adult tissue are induced to exit the growth cycle and enter the Go phase of the cell cycle before nuclear transfer.

Three new populations of cells were derived from (1) a day-9 embryo, (2) a day-26 fetus and (3) mammary gland of a 6-year-old ewe in the last trimester of pregnancy. Morphology of the embryo-derived cells is unlike both mouse embryonic stem (ES) cells and the embryo-derived cells used in our previous study. Nuclear transfer was carried out according to one of our established protocols[1] and reconstructed embryos transferred into recipient ewes. Ultrasound scanning detected 21 single fetuses on day 50–60 after oestrus. On subsequent scanning at ~ 14-day intervals, fewer fetuses were observed, suggesting either misdiagnosis or fetal loss. In total, 62% of fetuses were lost, a significantly greater proportion than the estimate of 6% after natural mating.[4] Increased prenatal loss has been reported after embryo manipulation or culture of unreconstructed embryos.[5] At about day 110 of pregnancy, four fetuses were dead, all from embryo-derived cells, and post-mortem analysis was possible after killing the ewes. Two fetuses had abnormal liver development, but no other abnormalities were detected and there was no evidence of infection.

Eight ewes gave birth to live lambs. All three cell populations were represented. One weak lamb, derived from the fetal fibroblasts, weighed 3.1 kg and died within a few minutes of birth, although post-mortem analysis failed to find any abnormality or infection. At 12.5%, perinatal loss was not dissimilar to that occurring in a large study of commercial sheep, when 8% of lambs died within 24 h of birth.[6] In all cases the lambs displayed the morphological characteristics of the breed used to derive the nucleus donors and not that of the oocyte donor. This alone indicates that the lambs could not have been born after inadvertent mating of either the oocyte donor or recipient ewes. In addition, DNA microsatellite analysis of the cell populations and the lambs at four polymorphic loci confirmed that each lamb was derived from the cell population used as nuclear donor. Duration of gestation is determined by fetal genotype,[7] and in all cases gestation was longer than the breed mean. By contrast, birth weight is influenced by both maternal and fetal genotype.[8] The birth weight of all lambs was within the range for single lambs born to Blackface ewes on our farm (up to 6.6 kg) and in most cases was within the range for the breed of the nuclear donor. There are no strict control observations for birth weight after embryo transfer between breeds, but the range in weight of lambs born to their own breed on our farms is 1.2–5.0 kg, 2–4.9 kg and 3–9 kg for the Finn Dorset, Welsh Mountain and Poll Dorset genotypes, respectively. The attainment of sexual maturity in the lambs is being monitored.

Development of embryos produced by nuclear transfer depends upon the maintenance of normal ploidy and creating the conditions for developmental regulation of gene expression. These responses are both influenced by the cell-cycle stage of donor and recipient cells and the interaction between them (reviewed in ref. 9). A comparison of development of mouse and cattle embryos produced by nuclear transfer to oocytes[10,11] or enucleated zygotes[12,13] suggests that a greater proportion develop if the recipient is an oocyte. This may be because factors that bring about reprogramming of gene expression in a transferred nucleus are required for early development and are taken up by the pronuclei during development of the zygote.

If the recipient cytoplasm is prepared by enucleation of an

oocyte at metaphase II, it is only possible to avoid chromosomal damage and maintain normal ploidy by transfer of diploid nuclei,[14,15] but further experiments are required to define the optimum cell-cycle stage. Our studies with cultured cells suggest that there is an advantage if cells are quiescent (ref. 1, and this work). In earlier studies, donor cells were embryonic blastomeres that had not been induced into quiescence. Comparisons of the phases of the growth cycle showed that development was greater if donor cells were in mitosis[16] or in the G1 (ref. 10) phase of the cycle, rather than in S or G2 phases. Increased development using donor cells in G0, G1 or mitosis may reflect greater access for reprogramming factors present in the oocyte cycoplasm, but a direct comparison of these phases in the same cell population is required for a clearer understanding of the underlying mechanisms.

Together these results indicate that nuclei from a wide range of cell types should prove to be totipotent after enhancing opportunities for reprogramming by using appropriate combinations of these cell-cycle stages. In turn, the dissemination of the genetic improvement obtained within elite selection herds will be enhanced by limited replication of animals with proven performance by nuclear transfer from cells derived from adult animals. In addition, gene targeting in livestock should now be feasible by nuclear transfer from modified cell populations and will offer new opportunities in biotechnology. The techniques described also offer an opportunity to study the possible persistence and impact of epigenetic changes, such as imprinting and telomere shortening, which are known to occur in somatic cells during development and senescence, respectively.

The lamb born after nuclear transfer from a mammary gland cell is, to our knowledge, the first mammal to develop from a cell derived from an adult tissue. The phenotype of the donor cell is unknown. The primary culture contains mainly mammary epithelial (over 90%) as well as other differentiated cell types, including myoepithelial cells and fibroblasts. We cannot exclude the possibility that there is a small proportion of relatively undifferentiated stem cells able to support regeneration of the mammary gland during pregnancy. Birth of the lamb shows that during the development of that mammary cell there was no irreversible modification of genetic information required for development to term. This is consistent with the generally accepted view that mammalian differentiation is almost all achieved by systematic, sequential changes in gene expression brought about by interactions between the nucleus and the changing cytoplasmic environment.[17]

Methods

Embryo-derived cells were obtained from embryonic disc of a day-9 embryo from a Poll Dorset ewe cultured as described,[1] with the following modifications. Stem-cell medium was supplemented with bovine DIA/LIF. After 8 days, the explanted disc was disaggregated by enzymatic digestion and cells replated onto fresh feeders. After a further 7 days, a single colony of large flattened cells was isolated and grown further in the absence of feeder cells. At passage 8, the modal chromosome number was 54. These cells were used as nuclear donors at passages 7–9. Fetal-derived cells were obtained from an eviscerated Black Welsh Mountain fetus recovered at autopsy on day 26 of pregnancy. The head was removed before tissues were cut into small pieces and the cells dispersed by exposure to trypsin. Culture was in BHK 21 (Glasgow MEM; Gibco Life Sciences) supplemented with L-glutamine (2 mM), sodium pyruvate (1 mM) and 10% fetal calf serum. At 90% confluency, the cells were passaged with a 1:2 division. At passage 4, these fibroblast-like cells had modal chromosome number of 54. Fetal cells were used as nuclear donors at passages 4–6. Cells from mammary gland were obtained from a 6-year-old Finn Dorset ewe in the last trimester of pregnancy.[18] At passages 3 and 6, the modal chromosome number was 54 and these cells were used as nuclear donors at passage numbers 3–6.

Nuclear transfer was done according to a previous protocol.[1] Oocytes were recovered from Scottish Blackface ewes between 28 and 33 h after injection of gonadotropin-releasing hormone (GnRH), and enucleated as soon as possible. They were recovered in calcium- and magnesium-free PBS containing 1% FCS and transferred to calcium-free M2 medium[19] containing 10% FCS at 37°C. Quiescent, diploid donor cells were produced by reducing the concentration of serum in the medium from 10 to 0.5% for 5 days, causing the cells to exit the growth cycle and arrest in G0. Confirmation that cells had left the cycle was obtained by staining with antiPCNA/cyclin antibody (Immuno Concepts), revealed by a second antibody conjugated with rhodamine (Dakopatts).

Fusion of the donor cell to the enucleated oocyte and activation of the oocyte were induced by the same electrical pulses, between 34 and 36 h after GnRH injection to donor ewes. The majority of reconstructed embryos were cultured in ligated oviducts of sheep as before, but some embryos produced by transfer from embryo-derived cells or fetal fibroblasts were cultured in a chemically defined medium.[20] Most embryos that developed to morula or blastocyst after 6 days of culture were transferred to recipients and allowed to develop to term. One, two or three embryos were transferred to each ewe depending upon the availability of embryos. The effect of cell type upon fusion and development to morula or blastocyst was analysed using the marginal model of Breslow and Clayton.[21] No comparison was possible of development to term as it was not practicable to transfer all embryos developing to a suitable stage for transfer. When too many embryos were available, those having better morphology were selected.

Ultrasound scan was used for pregnancy diagnosis at around day 60 after oestrus and to monitor fetal development thereafter at 2-week intervals. Pregnant recipient ewes were monitored for nutritional status, body condition and signs of EAE, Q fever, border disease, louping ill and toxoplasmosis. As lambing approached, they were under constant observation and a veterinary surgeon called at the onset of parturition. Microsatellite analysis was carried out on DNA from the lambs and recipient ewes using four polymorphic ovine markers.[22]

References

1. Campbell, K. H. S., McWhir, J., Ritchie, W. A. & Wilmut, I. "Sheep cloned by nuclear transfer from a cultured cell line." *Nature* **380,** 64–66 (1996).

2. Solter, D. "Lambing by nuclear transfer." *Nature* **380,** 24–25 (1996).

3. Gurdon, J. B., Laskey, R. A. & Reeves, O. R. "The developmental capacity of nuclei transplanted from keratinized skin cells of adult frogs." *J. Embryol. Exp. Morph.* **34,** 93–112 (1975).

4. Quinlivan, T. D., Martin, C. A., Taylor, W. B. & Cairney, I. M. "Pre- and perinatal mortality in those ewes that conceived to one service." *J. Reprod. Fertil.* **11,** 379–390 (1966).

5. Walker, S. K., Heard, T. M. & Seamark, R. F. "*In vitro* culture of sheep embryos without co-culture: successes and perspectives." *Therio* **37,** 111–126 (1992).

6. Nash, M. L., Hungerford, L. L., Nash, T. G. & Zinn, G. M. "Risk factors for perinatal and postnatal mortality in lambs." *Vet. Rec.* **139,** 64–67 (1996).

7. Bradford, G. E., Hart, R., Quirke, J. F. & Land, R. B. "Genetic control of the duration of gestation in sheep." *J. Reprod. Fertil.* **30,** 459–463 (1972).

8. Walton, A. & Hammond, J. "The maternal effects on growth and conformation in Shire horse–Shetland pony crosses." *Proc. R. Soc. B* **125,** 311–335 (1938).

9. Campbell, K. H. S., Loi, P., Otaegui, P. J. & Wilmut, I. "Cell cycle co-ordination in embryo cloning by nuclear transfer." *Rev. Reprod.* **1,** 40–46 (1996).

10. Cheong, H.-T., Takahashi, Y. & Kanagawa, H. "Birth of mice after transplantation of early-cell-cycle-stage embryonic nuclei into enucleated oocytes." *Biol. Reprod.* **48,** 958–963 (1993).

11. Prather, R. S. *et al.* "Nuclear transplantation in the bovine embryo. Assessment of donor nuclei and recipient oocyte." *Biol. Reprod.* **37,** 859–866 (1987).

12. McGrath, J. & Solter, D. "Inability of mouse blastomere nuclei transferred to enucleated zygotes to support development *in vitro.*" *Science* **226,** 1317–1318 (1984).

13. Robl, J. M. *et al.* "Nuclear transplantation in bovine embryos." *J. Anim. Sci.* **64,** 642–647 (1987).

14. Campbell, K. H. S., Ritchie, W. A. & Wilmut, I. "Nuclear-cytoplasmic interactions during the first cell cycle of nuclear transfer reconstructed bovine embryos: Implications for deoxyribonucleic acid replication and development." Biol. Reprod. 49, 933–942 (1993).

15. Barnes, F. L. et al. "Influence of recipient oocyte cell cycle stage on DNA synthesis, nuclear envelope breakdown, chromosome constitution, and development in nuclear transplant bovine embryos." Mol. Reprod. Dev. 36, 33–41 (1993).

16. Kwon, O. Y. & Kono, T. "Production of identical sextuplet mice by transferring metaphase nuclei from 4-cell embryos." J. Reprod. Fertil. Abst. Ser. 17, 30 (1996).

17. Gurdon, J. B. The Control of Gene Expression in Animal Development (Oxford University Press, Oxford, 1974).

18. Finch, L. M. B. et al. "Primary culture of ovine mammary epithelial cells." Biochem. Soc. Trans. 24, 369S (1996).

19. Whitten, W. K. & Biggers, J. D. "Complete development in vitro of the preimplantation stages of the mouse in a simple chemically defined medium." J. Reprod. Fertil. 17, 399–401 (1968).

20. Gardner, D. K., Lane, M., Spitzer, A. & Batt, P. A. "Enhanced rates of cleavage and development for sheep zygotes cultured to the blastocyst stage in vitro in the absence of serum and somatic cells. Amino acids, vitamins, and culturing embryos in groups stimulate development." Biol. Reprod. 50, 390–400 (1994).

21. Breslow, N. E. & Clayton, D. G. "Approximate inference in generalized linear mixed models." J. Am. Stat. Assoc. 88, 9–25 (1993).

22. Buchanan, F. C., Littlejohn, R. P., Galloway, S. M. & Crawford, A. L. "Microsatellites and associated repetitive elements in the sheep genome." Mammal. Gen. 4, 258–264 (1993).

We thank A. Colman for his involvement throughout this experiment and for guidance during the preparation of this manuscript; C. Wilde for mammary-derived cells; M. Ritchie, J. Bracken, M. Malcolm-Smith, W. A. Ritchie, P. Ferrier and K. Mycock for technical assistance; D. Waddington for statistical analysis; and H. Bowran and his colleagues for care of the animals. This research was supported in part by the Ministry of Agriculture, Fisheries and Food. The experiments were conducted under the Animals (Scientific Procedures) Act 1986 and with the approval of the Roslin Institute Animal Welfare and Experiments Committee.

This article first appeared in Nature, volume 385, February 27, 1997, pages 810–813. The figures and tables originally accompanying the article can be found there; they have been deleted for this book.

The Science and Application of Cloning

National Bioethics Advisory Commission

What Is Cloning?

*T*he word *clone* is used in many different contexts in biological research but in its most simple and strict sense, it refers to a precise genetic copy of a molecule, cell, plant, animal, or human being. In some of these contexts, *cloning* refers to established technologies that have been part of agricultural practice for a very long time and currently form an important part of the foundations of modern biological research.

Indeed, genetically identical copies of whole organisms are commonplace in the plant breeding world and are commonly referred to as *varieties* rather than *clones*. Many valuable horticultural or agricultural strains are maintained solely by vegetative propagation from an original plant, reflecting the ease with which it is possible to regenerate a complete plant from a small cutting. The developmental process in animals does not usually permit cloning as easily as in plants. Many simpler invertebrate species, however, such as certain kinds of worms, are capable of regenerating a whole organism from a small piece, even though this is not necessarily their usual mode of reproduction. Vertebrates have lost this ability entirely, although regeneration of certain limbs, organs, or tissues can occur to varying degrees in some animals.

Although a single adult vertebrate cannot generate another whole organism, cloning of vertebrates does occur in nature, in a limited way, through multiple births, primarily with the formation of identical twins. However, twins occur by chance in humans and other mammals, with the separation of a single embryo into halves at an early stage of development. The resulting offspring are genetically identical, having been derived from one zygote, which resulted from the fertilization of one egg by one sperm.

At the molecular and cellular level, scientists have been cloning human and animal cells and genes for several decades. The scientific justification for such cloning is that it provides greater quantities of identical cells or genes for study; each cell or molecule is identical to the others.

At the simplest level, molecular biologists routinely make clones of deoxyribonucleic acid (DNA), the molecular basis of genes. DNA fragments containing genes are copied and amplified in a host cell, usually a bacterium. The availability of large quantities of identical DNA makes possible many scientific experiments. This process, often called *molecular cloning,* is the mainstay of recombinant DNA technology and has led to the production of such important medicines as insulin to treat diabetes, tissue plasminogen activator (tPA) to dissolve clots after a heart attack, and erythropoietin (EPO) to treat anemia associated with dialysis for kidney disease.

Another type of cloning is conducted at the cellular level. In *cellular cloning* copies are made of cells derived from the soma, or body, by growing these cells in culture in a laboratory. The genetic makeup of the resulting cloned cells, called a *cell line,* is identical to that of the original cell. This, too, is a highly reliable procedure, which is also used to test and sometimes to produce new medicines such as those listed above. Since molecular and cellular cloning of this sort does not involve germ cells (eggs or sperm), the cloned cells are not capable of developing into a baby.

In the early 1980s, a more sophisticated form of cloning animals was developed, known as *nuclear transplantation cloning*. The nucleus of somatic cells is diploid—that is, it contains two sets of genes, one from the mother and one from the father. Germ cells, however, contain a haploid nucleus, with only the maternal or paternal genes. In nuclear transplantation cloning, the nucleus is removed from an egg and replaced with the diploid nucleus of a somatic cell. In such nuclear transplantation cloning there is a single genetic "parent," unlike sexual reproduction where a new organism is formed when the genetic material of the egg and sperm fuse. The first experiments of this type were successful only when the donor cell was derived from an early embryo. In theory, large numbers of genetically identical animals could be produced through such nuclear transplantation cloning. In practice, the nuclei from embryos which have developed beyond a certain number of cells seem to lose their totipotency, limiting the number of animals that can be produced in a given period of time from a single, originating embryo.

The new development in the experiments that Wilmut and colleagues carried out to produce Dolly was the use of much more developed somatic cells isolated from adult sheep as the source of the donor nuclei. This achievement of gestation and live birth of a sheep using an adult cell donor nucleus was stunning evidence that cell differentiation and specialization are reversible. Given the fact that cells develop and divide after fertilization and differentiate into specific tissue (e.g., muscle, bone, neurons), the development of a viable adult sheep from a differentiated adult cell nucleus provided surprising evidence that the pattern of gene expression can be reprogrammed. Until this experiment many biologists believed that reactivation of the genetic material of mammalian somatic cells would not be complete enough to allow for the production of a viable adult mammal from nuclear transfer cloning.

Remaining Scientific Uncertainties

Several important questions remain unanswered about the feasibility in mammals of nuclear transfer cloning using adult cells as the source of nuclei:

First, can the procedure that produced Dolly be carried out successfully in other cases? Only one animal has been produced to date. Thus, it is not clear that this technique is reproducible even in sheep.

The successful generation of an adult sheep from a somatic cell nucleus suggests that the imprint can be stable, but it is possible that some instability of the imprint, particularly in cells in culture, could limit the efficiency of nuclear transfer from somatic cells. It is known that disturbances in imprinting lead to growth abnormalities in mice and are associated with cancer and rare genetic conditions in children.

[W]ill the mutations that accumulate in somatic cells affect nuclear transfer efficiency and lead to cancer and other diseases in the offspring? As cells divide and organisms age, mistakes and alterations (mutations) in the DNA will inevitably occur and will accumulate with time. If these mistakes occur in the sperm or the egg, the mutation will be inherited in the offspring. Normally mutations that occur in somatic cells affect only that cell and its descendants which are ultimately dispensable. Nevertheless, such mutations are not necessarily harmless. Sporadic somatic mutations in a variety of genes can predispose a cell to become cancerous. Transfer of a nucleus from a somatic cell carrying such a mutation into an egg would transform a sporadic somatic mutation into a germline mutation that is transmitted to all of the cells of the body. If this mutation were present in all cells it may lead to a genetic disease or cancer. The risks of such events occurring following nuclear transfer are difficult to estimate.

Why Pursue Animal Cloning Research?

Research on nuclear transfer cloning in animals may provide information that will be useful in biotechnology, medicine, and basic science. Some of the immediate goals of this research are:

- to generate groups of genetically identical animals for research purposes
- to rapidly propagate desirable animal stocks
- to improve the efficiency of generating and propagating transgenic livestock
- to produce targeted genetic alterations in domestic animals
- to pursue basic knowledge about cell differentiation

CLONING ANIMALS FOR RESEARCH PURPOSES

Inbred strains of mice have been a mainstay of biological research for years because they are essentially genetically identical and homozygous (i.e., both copies of each gene inherited from the mother and father are identical). Experimental analysis is simplified because differences in genetic background that often lead to experimental variation are eliminated. Generating such homozygous inbred lines in larger animals is difficult and time-consuming because of the long gestation times and small numbers of offspring.

Repeated cycles of nuclear transfer can expand the number of individual animals derived from one donor nucleus, allowing more identical animals to be generated. The first nuclear transfer embryo is allowed to divide to early blastomere stages and then those cells are used as donor nuclei for another series of transfers. This process can be carried on indefinitely, in theory, although practice suggests that successful fusion rates decline with each cycle of transfer. One experiment in cows, for example, produced 54 early embryos after three cycles of transfer from a single blastomere nucleus from one initial embryo (Stice and Keefer, 1993). Viable calves were produced from all three cycles of nuclear transfer.

This approach is likely to be limited in its usefulness, however. A group of cloned animals derived from nuclear transfer from an individual animal is self-limited. Unless they are derived from an inbred stock initially, each clone derived from one individual will differ genetically from a clone derived from another individual. Once a cloned animal is mated to produce offspring, the offspring will no longer be identical due to the natural processes which shuffle or recombine genes during development of eggs and sperm. Thus, each member of a clone has to be made for each experiment by nuclear transfer, and generation of a large enough number of cloned animals to be useful as experimental groups is likely to be prohibitively expensive in most animals.

ADVANTAGES OF NUCLEAR TRANSFER CLONING FOR BREEDING LIVESTOCK

Nuclear transfer cloning, especially from somatic cell nuclei, could provide an additional means of expanding the number of chosen livestock. The ability to make identical copies of adult prize cows, sheep, and pigs is a feature unique to nuclear transfer technologies, and may well be used in livestock production, if the efficiencies of adult nuclear transfer can be improved. The net effect of multiplying chosen animals by cloning will be to reduce the overall genetic diversity in a given livestock line, likely with severe adverse long-term consequences. If this technique became widespread, efforts would have to be made to ensure a pool of genetically diverse animals for future livestock maintenance.

IMPROVED GENERATION AND PROPAGATION OF TRANSGENIC LIVESTOCK

There is considerable interest in being able to genetically alter farm animals by introduction and expression of genes from other species, such as humans. So-called "transgenic animals" were first developed using mice, by microinjection of DNA into the nucleus of the egg. This ability to add genes to an organism has been a major research tool for understanding gene regulation and for using the mouse as a model in studies of certain human diseases. It has also been applied to other species including livestock.

Currently, the major activity in livestock transgenesis is focused on pharmaceutical and medical applications. The milk of livestock animals can be modified to contain large amounts of pharmaceutically important proteins such as insulin or factor VIII for treatment of human disease by expressing human genes in the mammary gland (Houdebine, 1994). In sheep greater than 50 percent of the proteins in milk can be the product of a human genes (Colman, 1996).

Another major area of interest is the use of transgenic animals for organ transplantation into humans. Pig organs, in many cases, are similar enough to humans to be potentially useful in organ transplants, if problems of rejection could be overcome.

Foreign DNA, such as a human gene, could be introduced into cell lines in culture and cells expressing the transgene could be characterized and used as a source of donor nuclei for cloning, and all offspring would likely express the human gene. This, in fact, was the motivation behind the experiments that led to the production of Dolly. If a human gene such as that for insulin could be expressed in the mammary gland, the milk of the sheep would be an excellent source of insulin to treat diabetes.

GENERATING TARGETED GENE ALTERATIONS

The most powerful technology for gene replacement in mammals was developed in mice. This technique adds manipulated or foreign DNA to cells in culture to replace the DNA present in the genome of the cells. Thus, mutations or other alterations that would be useful in medical research can be introduced into an animal in a directed and controlled manner and their effects studied, a process called *gene targeting* (Capecchi, 1989).

This use of gene replacement has been responsible for the explosion in the generation of "knock-out" mice, in which specific genes have been deleted from the genome. These mice have been invaluable in current studies to understand normal gene function and to allow the generation of accurate models of human genetic disease. Gene targeting approaches can also be used to ensure correct tissue-specific expression of foreign genes and to suppress the expression of genes in inappropriate tissues. If applied to domestic animals, this technology could increase the efficiency of the expression of foreign genes by targeting the introduced genes to appropriate regions of the chromosome. It could also be used to directly alter the normal genes of the organism, which could influence animal health and productivity or to help develop transgenic organs that are less likely to be rejected upon transplantation.

BASIC RESEARCH ON CELL DIFFERENTIATION

The basic cellular processes that allowed the birth of Dolly by nuclear transfer using the nucleus from an adult somatic donor cell are not well understood. If indeed the donor cell was a fully differentiated cell and not a rare, less differentiated stem cell that resulted in this cloned sheep, there will be many questions to ask about how this process occurred. How the specialized cell from the mammary gland was reprogrammed to allow the expression of a complete developmental program will be a fascinating area of study. Developmental biologists will want to know which genes are reprogrammed, when they are expressed, and in what order. This might shed light on the still poorly understood process of sequential specialization that must occur during development of all organisms.

Basic research into these fundamental processes may also lead to the development of new therapies to treat human disease. It is not possible to predict from where the essential new discoveries will come. However, already the birth of Dolly has sparked ideas about potential benefits that might be realized. To explore the possibility of these new therapies, extensive basic research is needed.

Potential Therapeutic Applications of Nuclear Transfer Cloning

The demonstration that, in mammals as in frogs, the nucleus of a somatic cell can be reprogrammed by the egg environment provides further impetus to studies on how to reactivate embryonic programs of development in adult cells. These studies have exciting prospects for regeneration and repair of diseased or damaged human tissues and organs, and may provide clues as to how to reprogram adult differentiated cells directly without the need for oocyte fusion. In addition, the use of nuclear transfer has potential application in the field of assisted reproduction.

POTENTIAL APPLICATIONS IN ORGAN AND TISSUE TRANSPLANTATION

Many human diseases, when they are severe enough, are treated effectively by organ or tissue transplantation, including some leukemias, liver failure, heart and kidney disease. In some instances the organ required is non-vital, that is, it can be taken from the donor without great risk (e.g., bone marrow, blood, kidney). In other cases, the organ is obviously vital and required for the survival of the individual, such as the heart. All transplantation is imperfect, with the exception of that which occurs between identical twins, because transplantation of organs between individuals requires genetic compatibility.

In principle, the application of nuclear transfer cloning to humans could provide a potential source of organs or tissues of a predetermined genetic background. The notion of using human cloning to produce individuals for use solely as organ donors is repugnant, almost unimaginable, and morally unacceptable. A morally more acceptable and potentially feasible approach is to direct differentiation along a specific path to produce specific tissues (e.g., muscle or nerve) for therapeutic transplantation rather than to produce an entire individual. Given current uncertainties about the feasibility of this, however, much research would be needed in animal systems before it would be scientifically sound, and therefore potentially morally acceptable, to go forward with this approach.

POTENTIAL APPLICATIONS IN CELL-BASED THERAPIES

Another possibility raised by cloning is transplantation of cells or tissues not from an individual donor but from an early embryo or embryonic stem cells, the primitive, undifferentiated cells from the embryo that are still totipotent. This potential application would not require the generation and birth of a cloned individual.

ASSISTED REPRODUCTION

Another area of medicine where the knowledge gained from animal work has potential application is in the area of assisted reproduction. Assisted reproduction technologies are already widely used and encompass a variety of parental and biological situations, that is, donor and recipient relationships. In most cases, an infertile couple seeks remedy through either artificial insemination or in vitro fertilization using sperm from either the male or an anonymous donor, an egg from the woman or a donor, and in some cases surrogacy. In those instances where both individuals of a couple are infertile or the prospective father has nonfunctional sperm, one might envision using cloning of one member of the couple's nuclei to produce a child.

Cloning and Genetic Determinism

The announcement of Dolly sparked widespread speculation about a human child being created using somatic cell nuclear transfer. Much of the perceived fear that greeted this announcement centered around the misperception that a child or many children could be produced who would be identical to an already existing person.

This fear reflects an erroneous belief that a person's genes bear a simple relationship to the physical and psychological traits that compose that individual. This belief, that genes alone determine all aspects of an individual, is called *genetic determinism*. Although genes play an essential role in the formation of physical and behavioral characteristics, each individual is, in fact, the result of a complex interaction between his or her genes and the environment within which they develop, beginning at the time of fertilization and continuing throughout life. As social and biological beings we are creatures of our biological, physical, social, political, historical, and psychological environments. Indeed, the great lesson of modern molecular genetics is the profound complexity of both gene-gene interactions and gene-environment interactions in the determination of whether a specific trait or characteristic is expressed. In other words, there will never be another you.

Thus, the idea that one could make through somatic cell nuclear transfer a team of Michael Jordans, a physics department of Albert Einsteins, or an opera chorus of Pavarottis, is simply false. Knowing the complete genetic makeup of an individual would not tell you what kind of person that individual would become. Even identical twins who grow up together and thus share the same genes and a similar home environment have different likes and dislikes, and can have very different talents. The increasingly sophisticated studies coming out of human genetics research are showing that the better we understand gene function, the less likely it is we will ever be able to produce at will a person with any given complex trait.

References

Capecchi, M. R., "The new mouse genetics: altering the genome by gene targeting," *Trends in Genetics* 5:70–76, 1989.

Colman, A., "Production of proteins in the milk of transgenic livestock: Problems, solutions, and successes," *American Journal of Clinical Nutrition* 63:639S–645S, 1996.

Houdebine, L. M., "Production of pharmaceutical proteins from transgenic animals," *Journal of Biotechnology* 34:269–287, 1994.

Stice, S. L., and C. L. Keefer, "Multiple generational bovine embryo cloning," *Biology of Reproduction* 48:715–719, 1993.

This is an abridged version of chapter 2 of the NBAC report on human cloning (June 1997). Much of the material in the original version is derived from two commissioned papers provided by Janet Rossant, Samuel Lunenfeld Research Institute, Mount Sinai Hospital, Toronto; and by Stuart Orkin, Children's Hospital and Dana Farber Cancer Institute, Boston.

Dolly's Fashion
and Louis's Passion

Stephen Jay Gould

Nothing can be more fleeting or capricious than fashion. What, then, can a scientist, committed to objective description and analysis, do with such a haphazardly moving target? In a classic approach, analogous to standard advice for preventing the spread of an evil agent ("kill it before it multiplies"), a scientist might say, "Quantify it before it disappears."

Francis Galton, Charles Darwin's charmingly eccentric and brilliant cousin, and a founder of the science of statistics, surely took this prescription to heart. He once decided to measure the geographic patterning of female beauty. He attached a piece of paper to a small wooden cross that he could carry, unobserved, in his pocket. He held the cross at one end in the palm of his hand and, with a needle secured between thumb and forefinger, made pinpricks on the three remaining projections (the two ends of the cross bar and the top). He would rank every young woman he passed on the street into one of three categories, as beautiful, average, or substandard (by his admittedly subjective preferences)—and he would then place a pinprick for each woman into the designated domain of his cross. After a hard day's work, he tabulated the relative percentages by counting pinpricks. He concluded, to the dismay of Scotland, that beauty followed a simple trend from north to south, with the highest proportion of uglies in Aberdeen, and the greatest frequency of lovelies in London.

Some fashions (tongue piercings, perhaps?) flower once and then disappear, hopefully forever. Others swing in and out of style, as if fastened to the end of a pendulum. Two foibles of human life strongly promote this oscillatory mode. First, our need to create order in a complex world begets our worst mental habit: dichotomy, or our tendency to reduce a truly intricate set of subtle shadings to a choice between two diametrically opposed alternatives (each with moral weight and therefore ripe for bombast and pontification, if not outright warfare): religion *vs.* science, liberal *vs.* conservative, plain *vs.* fancy, "Roll Over Beethoven" *vs.* the *Moonlight* Sonata. Second, many deep questions about our loves and livelihoods, and the fates of nations, truly have no answers—so we cycle the presumed alternatives of our dichotomies, one after the other, always hoping that, this time, we will find the nonexistent key.

Among oscillating fashions governed primarily by the swing of our social pendulum, no issue could be more prominent for an evolutionary biologist, or more central to a broad range of political questions, than genetic *vs.* environmental sources of human abilities and behaviors. This issue has been falsely dichotomized for so many centuries that English even features a mellifluous linguistic contrast for the supposed alternatives: nature *vs.* nurture.

As any thoughtful person understands, the framing of this question as an either-or dichotomy verges on the nonsensical. Both inheritance and upbringing matter in crucial ways. Moreover, an adult human being, built by interaction of these (and other) factors, cannot be disaggregated into separate components with attached percentages. It behooves us all to grasp why such common claims as "intelligence is 30 percent genetic and 70 percent environmental" have no sensible meaning at all, and represent the same kind of error as the contention that all overt properties of water may be revealed by noting an underlying construction as two parts of one gas mixed with one part of another.

Nonetheless, a preference for either nature or nurture swings back and forth into fashion as political winds blow, and as scientific breakthroughs grant transient prominence to one or another feature in a spectrum of vital influences. For example, a combination of political and scientific factors favored an emphasis upon environment in the years just following World War II: an understanding that Hitlerian horrors had been rationalized by claptrap genetic theories about inferior races; the heyday of behaviorism in psychology. Today, genetic explanations are all the rage, fostered by a similar mixture of social and scientific influences: for example, the rightward shift of the political pendulum (and the cynical availability of "you can't change them, they're made that way" as a bogus argument for reducing government expenditures on social programs); an overextension to all behavioral variation of genuinely exciting results in identifying the genetic basis of specific diseases, both physical and mental.

Unfortunately, in the heat of immediate enthusiasm, we often mistake transient fashion for permanent enlightenment. Thus, many people assume that the current popularity of genetic determinism represents a final truth wrested from the clutches of benighted environmentalists of previous generations. But the lessons of history suggest that the worm will soon turn again. Since both nature and nurture can teach us so much—and since the fullness of our behavior and mentality represents such a complex and unbreakable combination of these and other factors—a current emphasis on nature will no doubt yield to a future fascination with nurture as we move towards better understanding by lurching upward from one side to another in our quest to fulfill the Socratic injunction: know thyself.

In my Galtonian desire to measure the extent of current fascination with genetic explanations (before the pendulum swings once again and my opportunity evaporates), I hasten to invoke two highly newsworthy items of recent months. The subjects may seem quite unrelated—Dolly the cloned sheep, and Frank Sulloway's book on the effects of birth order upon human behavior—but both stories share a curious common feature offering striking insight into the current extent of genetic preferences. In short, both stories have been reported almost entirely in genetic terms, but both cry out (at least to me) for a reading as proof of strong environmental influences. Yet no one seems to be drawing (or even mentioning) this glaringly obvious inference. I cannot imagine that anything beyond current fashion for genetic arguments can explain this puzzling silence. I am convinced that exactly the same information, if presented twenty years ago in a climate favoring explanations based on nurture, would have been read primarily in this opposite light. Our world, beset by ignorance and human nastiness, contains quite enough background darkness. Should we not let both beacons shine all the time?

Creating Sheep

Dolly must be the most famous sheep since John the Baptist designated Jesus in metaphor as "lamb of God, which taketh away the sins of the world" (John 1:29). She has certainly edged past the pope, the president, Madonna, and Michael Jordan as the best-known mammal of the moment. And all this for a carbon copy, a Xerox! I don't mean to drop cold water on this little lamb, cloned from a mammary cell of her adult mother, but I remain unsure that she's worth all the fuss and fear generated by her unconventional birth.

When one reads the technical article describing Dolly's manufacture (see previous essay, by Wilmut *et al.*), rather than the fumings and hyperbolae of so much public commentary, one can't help feeling a bit underwhelmed, and left wondering whether Dolly's story tells less than meets the eye.

I don't mean to discount or underplay the ethical issues raised by Dolly's birth (and I shall return to this subject in a moment), but we are not about to face an army of Hitlers or even a Kentucky Derby run entirely by genetically identical contestants (a true test for the skills of jockeys and trainers)! First, Dolly breaks no theoretical ground in biology, for we have known how to clone in principle for at least two decades, but had developed no techniques for reviving the full genetic potential of differentiated adult cells. (Still, I admit that a technological solution can pack as much practical and ethical punch as a theoretical breakthrough. I suppose one could argue that the first atomic bomb only realized a known possibility.)

Second, my colleagues have been able to clone animals from embryonic cells lines for several years, so Dolly is not the first mammalian clone, but only the first clone from an adult cell. Wilmut and colleagues also cloned sheep from cells of a 9-day embryo and a 26-day fetus—and had much greater success. They achieved 15 pregnancies (though not all proceeded to term) in 32 "recipients" (that is, surrogate mothers for transplanted cells) of the embryonic cell-line, 5 pregnancies in 16 recipients of the fetal cell-line, but only Dolly (1 pregnancy in 13 tries) for the adult cell-line. This experiment cries out for confirming repetition. (Still, I allow that current difficulties will surely be overcome, and cloning from adult cells, if doable at all, will no doubt be achieved more routinely as techniques and familiarity improve.)

Third, and more seriously, I remain unconvinced that we should regard Dolly's starting cell as adult in the usual sense of the term. Dolly grew from a cell taken from the "mammary gland of a 6-year-old ewe in the last trimester of pregnancy" (to quote the technical article of Wilmut *et al.*). Since the breasts of pregnant mammals enlarge substantially in late stages of pregnancy, some mammary cells, though technically adult, may remain unusually labile or even "embryo like," and thus able to proliferate rapidly to produce new breast tissue at an appropriate stage of pregnancy. Consequently, we may only be able to clone from unusual adult cells with effectively embryonic potential, and not from any stray cheek cell, hair follicle, or drop of blood that happens to fall into the clutches of a mad xeroxer. Wilmut and colleagues admit this possibility in a sentence written with all the obtuseness of conventional scientific prose, and therefore almost universally missed by journalists: "We cannot exclude the possibility that there is a small proportion of relatively undifferentiated stem cells able to support regeneration of the mammary gland during pregnancy."

But if I remain relatively unimpressed by achievements thus far, I do not discount the monumental ethical issues raised by the possibility of cloning from adult cells. Yes, we have cloned fruit trees for decades by the ordinary process of grafting—and without raising any moral alarms. Yes, we may not face the evolutionary dangers of genetic uniformity in crop plants and livestock, for I trust that plant and animal breeders will not be stupid enough to eliminate all but one genotype from a species, and will always maintain (as plant breeders do now) an active pool of genetic diversity in reserve. (But then, I suppose we should never underestimate the potential extent of human stupidity—and localized reserves could be destroyed by a catastrophe, while genetic diversity spread throughout a species guarantees maximal evolutionary robustness.)

Nonetheless, while I regard many widely expressed fears as exaggerated, I do worry deeply about potential abuses of human cloning, and I do urge a most open and thorough debate on these issues. Each of us can devise a personal worst-case scenario. Somehow, I do not focus upon the spectre of a future Hitler making an army of 10 million identical robotic killers—for if our society ever reaches a state where such an outcome might be realized, we are probably already lost. My thoughts run to localized moral quagmires that we might actually have to face in the next few years—for example, the biotech equivalent of ambulance-chasing slimeballs among lawyers: a husting little firm that scans the orbits for reports of children who died young, and then goes to grieving parents with the following offer: "So sorry for your loss; but did you save a hair sample? We can make you another for a mere 50 thou."

However, and still on the subject of ethical conundrums, but now moving to my main point about current underplaying of environmental sources for human behaviors, I do think that the most potent scenarios of fear, and the most fretful ethical discussions on late-night television, have focused on a nonexistent problem that all human societies solved millennia ago. We ask: Is a clone an individual? Would a clone have a soul? Would a clone made from my cell negate my unique personhood?

May I suggest that these endless questions—all variations on the theme that clones threaten our traditional concept of individuality—have already been answered empirically, even though public discussion of Dolly seems blithely oblivious to this evident fact. We have known human clones from the dawn of our consciousness. We call them identical twins—and they are far better clones than Dolly and her mother. Dolly only shares nuclear DNA with her mother's mammary cell—for the nucleus of this cell was inserted into an embryonic stem-cell (whose own nucleus had been removed) of a surrogate female. Dolly then grew in the womb of this surrogate.

Identical twins share at least four additional (and important) attributes that differ between Dolly and her mother. First, identical twins also house the same mitochondrial genes. (Mitochondria, the "energy factories" of cells, contain a small number of genes. We get our mitochondria from the cytoplasm of the egg cell that made us, not from the nucleus formed by the union of sperm and egg. Dolly received her nucleus from her mother, but her egg cytoplasm, and hence her mitochondria, from her surrogate.) Second, identical twins share the same set of maternal gene products in the egg. Genes don't grow embryos all by themselves. Egg cells contain protein products of maternal genes that play a major role in directing the early development of the embryo. Dolly has her mother's nuclear genes, but her surrogate's gene products in the cytoplasm of her founding cell.

Third—and now we come to explicitly environmental factors—identical twins share the same womb. Dolly and her mother gestated in different places. Fourth, identical twins share the same time and culture (even if they fall into the rare category, so cherished by researchers, of siblings separated at birth and raised, unbeknownst to each other, in distant families of different social classes). The clone of an adult cell matures in a different world. Does anyone seriously believe that a clone of Beethoven would sit down one day to write a tenth symphony in the style of his early-nineteenth-century forebear?

So identical twins are truly eerie clones—ever so much more alike on all counts, than Dolly and her mother. We do know that identical twins share massive similarities, not only of appearance, but also in broad propensities and detailed quirks of personality. Nonetheless, have we ever doubted the personhood of each member in a pair of identical twins? Of course not. We know that identical twins are distinct individuals, albeit with peculiar and extensive similarities. We give them different names. They encounter divergent experiences and fates. Their lives wander along disparate paths of the world's complex vagaries. They grow up as distinctive and undoubted individuals, yet they stand forth as far better clones than Dolly and her mother.

Why have we overlooked this central principle in our fears about Dolly? Identical twins provide sturdy proof that inevitable differences of nurture guarantee the individuality and personhood of each human clone. And since any future human Dolly must differ far more from her progenitor (in both the nature of mitochondria and maternal gene products, and the nurture of different wombs and surrounding cultures) than any identical twin diverges from her sibling clone, why ask if Dolly has a soul or an independent life when we have never doubted the personhood or individuality of far more similar identical twins?

Literature has always recognized this principle. The Nazi loyalists who cloned Hitler in *The Boys from Brazil* also understood that they had to maximize similarities of nurture as well. So they fostered their little Hitler babies in families maximally like Adolph's own dysfunctional clan—and not one of them grew up anything like history's quintessential monster. Life has always verified this principle as well. Eng and Chang, the original Siamese twins and the closest clones of all, developed distinct and divergent personalities. One became a morose alcoholic, the other remained a benign and cheerful man. We may not think much of the individuality of sheep in general (for they do set our icon of blind following and identical form as they jump over fences in the mental schemes of insomniacs), but Dolly will grow up to be as unique and as ornery as any sheep can be.

Killing Kings

My friend Frank Sulloway recently published a book that he had fretted over, nurtured, massaged, and lovingly shepherded towards publication for more than two decades. Frank and I have been discussing his thesis ever since he began his studies. I thought (and suggested) that he should have published his results twenty years ago. I still hold this opinion—for, while I greatly admire his book, and do recognize that such a long gestation allowed Frank to strengthen his case by gathering and refining his data, I also believe that he became too committed to his central thesis, and tried to extend his explanatory umbrella over too wide a range, with arguments that sometimes smack of special pleading and tortured logic.

Born to Rebel documents a crucial effect of birth order in shaping human personalities and styles of thinking. Firstborns, as sole recipients of parental attention until the arrival of later children, and as more powerful (by virtue of age and size) than their subsequent siblings, tend to cast their lot with parental authority and with the advantages of incumbent strength. They tend to grow up competent and confident, but also conservative and unlikely to favor quirkiness or innovation. Why threaten an existing structure that has always offered you clear advantages over siblings? Later children, however, are (as Sulloway's title proclaims) born to rebel. They must compete against odds for parental attention long focused primarily elsewhere. They must scrap and struggle, and learn to make do for themselves. Laterborns therefore tend to be flexible, innovative and open to change. The business and political leaders of stable nations may be overwhelmingly firstborns, but the revolutionaries who have discombobulated our cultures and restructured our scientific knowledge tend to be laterborns.

Sulloway defends his thesis with statistical data on the relationship of birth order and professional achievement in modern societies, and by interpreting historical patterns as strongly influenced by characteristic differences in behaviors of firstborns and laterborns. I found some of his historical arguments fascinating and persuasive when applied to large samples (but often uncomfortably overinterpreted in attempts to explain the intricate details of individual lives, for example, the effect of birth order on the differential success of Henry VIII's various wives in overcoming his capricious cruelties).

In a fascinating case, Sulloway chronicles a consistent shift in relative percentages of firstborns among successive groups in power during the French Revolution. The moderates initially in charge tended to be firstborns. As the revolution became more radical, but still idealistic and open to innovation and free discussion, laterborns strongly predominated. But when control then passed to the uncompromising hardliners who promulgated the Reign of Terror, firstborns again ruled the roost. In a brilliant stroke, Sulloway tabulates the birth orders for several hundred delegates who decided the fate of Louis XVI in the National Convention. Among hardliners who voted for the guillotine, 73 percent were firstborns; but 62 percent of the delegates who opted for the compromise of conviction with pardon were laterborns. Since Louis lost his head by a margin of one vote, an ever so slightly different mix of birth orders among delegates might have altered the course of history.

Since Frank is a good friend, and since I have been at least a minor midwife to this project over two decades, I took an unusually strong interest in the delayed birth of *Born to Rebel.* I read the text and all the prominent reviews that appeared in many newspapers and journals. And I have been puzzled—stunned would not be too strong a word—by the total absence from all commentary of the simplest and most evident inference from Frank's data, the one glaringly obvious point that everyone should have stressed, given the long history of issues raised by such information.

Sulloway focuses nearly all his interpretation on an extended analogy (broadly valid in my judgement, but overextended as an exclusive device) between birth order in families and ecological status in a world of Darwinian competition. Children vie for limited parental resources, just as individuals struggle for existence (and ultimately for reproductive success) in nature. Birth orders place children in different "niches," requiring disparate modes of competition for maximal success. While firstborns shore up incumbent advantages, laterborns must grope and grub by all clever means at their disposal—leading to the divergent personalities of stalwart and rebel. Alan Wolfe, in my favorite negative review from the *New Republic* (December 2 3, 1996; Jared Diamond stresses the same themes in my favorite positive review from the *New York Review of Books,* November 1 4, 1996), writes: "Since firstborns already occupy their own niches, laterborns, if they are to be noticed, have to find unoccupied niches. If they do so successfully, they will be rewarded with parental investment."

As I said, I am willing to go with this program up to a point. But I must also note that the restriction of commentary to this Darwinian metaphor has diverted attention from the foremost conclusion revealed by a large effect of birth order upon human behavior. The Darwinian metaphor smacks of biology; we also erroneously think of biological explanations as intrinsically genetic (an analysis of this common fallacy could fill an essay or an entire book). I suppose that this chain of argument leads us to stress whatever we think that Sulloway's thesis might be teaching us about "nature" (our preference, in any case, during this age of transient fashion for genetic causes) under our erroneous tendency to treat the explanation of human behavior as a debate between nature and nurture.

But consider the meaning of birth-order effects for environmental influences, however unfashionable at the moment. Siblings differ genetically of course, but no aspect of this genetic variation correlates in any systematic way with birth order. Firstborns and laterborns receive the same genetic shake within a family. Systematic differences in behavior between firstborns and laterborns cannot be ascribed to genetics. (Other biological effects may correlate with birth order—if, for example, the environment of the womb changes systematically with numbers of pregnancies—but such putative influences have no basis in genetic differences among siblings.) Sulloway's substantial birth-order effects therefore provide our best and ultimate documentation of nurture's power. If birth order looms so large in setting the paths of history and the allocation of people to professions, then nurture cannot be denied a powerfully formative role in our intellectual and behavioral variation and in setting the paths of human history as well. To be sure, we often fail to see what stares us in the face; but how can the winds of fashion blow away such an obvious point, one so relevant to our deepest and most persistent questions about ourselves?

In this case, I am especially struck by the irony of fashion's veil. As noted before, I urged Sulloway to publish his data twenty years ago—when (in my judgement) he could have presented an even better case because he had already documented the strong and general influence of birth order upon personality, but had not yet ventured upon the slippery path of trying to explain too many details with forced arguments that sometimes lapse into self-parody. If Sulloway had published in the mid 1970s, when nurture rode the pendulum of fashion in a politically more liberal age (probably dominated by laterborns!), I am confident that this obvious point about birth-order effects as proof of nurture's power would have won primary attention, rather than consignment to a limbo of invisibility.

Hardly anything in intellectual life can be more salutatory than the separation of fashion from fact. Always suspect fashion (especially when the moment's custom matches your personal predilection); always cherish fact (while remembering that an apparent "fact" may only record a transient fashion). I have discussed two subjects that couldn't be "hotter," but cannot be adequately understood because a veil of genetic fashion now conceals the richness of full explanation by relegating a preeminent environmental theme to invisibility. Thus, we worry whether the first cloned sheep represents a genuine individual at all, while we forget that we have never doubted the distinct personhood guaranteed by differences in nurture to clones far more similar by nature than Dolly and her mother—identical twins. And we try to explain the strong effects of birth order only by invoking a Darwinian analogy between family place and ecological niche, while forgetting that these systematic effects cannot have a genetic basis, and therefore prove the predictable power of nurture. So, sorry Louis. You lost your head to the power of family environments upon head children. And hello Dolly. May we forever restrict your mode of manufacture, at least for humans. But may genetic custom never stale the infinite variety guaranteed by a lifetime of nurture in the intricate complexity of nature—this vale of tears, joy and endless wonder.

This essay first appeared in *Natural History*, vol. 106, no. 5, June 1997, pages 18–23, 76.

What's Wrong with Cloning?

Richard Dawkins

S*cience* and logic cannot tell us what is right and what is wrong (Dawkins, 1998). You cannot, as I was once challenged to do by a belligerent radio interviewer, prove logically from scientific evidence that murder is wrong. But you can deploy logical reasoning, and even scientific facts, in demonstrating to dogmatists that their convictions are mutually contradictory. You can prove that their passionate denunciation of X is incompatible with their equally passionate advocacy of Y, because X and Y, though they had not realized it before, are the same thing (Glover, 1984). Science can show us a new way of thinking about an issue, perhaps open our imaginations in unexpected ways, with the consequence that we see our personal Xs and Ys in different ways and our values change. Sometimes we can be shown a way of seeing that makes us feel more favorably disposed to something that had been distasteful or frightening. But we can also be alerted to menacing implications of something that we had previously thought harmless or frivolously amusing. Cloning provides a case study in the power of scientific thinking to change our minds, in both directions.

Public responses to Dolly the sheep varied but, from President Clinton down, there was almost universal agreement that such a thing must never be allowed to happen to humans. Even those arguing for the medical benefits of cloning human tissues in culture were careful to establish their decent credentials, in the most vigorous terms, by denouncing the very thought that adult humans might be cloned to make babies, like Dolly.

But is it so obviously repugnant that we shouldn't even think about it? Mightn't even you, in your heart of hearts, quite like to be cloned? As Darwin said in another context, it is like confessing a murder, but I think I would. The motivation need have nothing to do with vanity, with thinking that the world would be a better place if there was another one of you living on after you are dead. I have no such illusions. My feeling is founded on pure curiosity. I know how I turned out, having been born in the 1940s, schooled in the 1950s, come of age in the 1960s, and so on. I find it a personally riveting thought that I could watch a small copy of myself, fifty years younger and wearing a baseball hat instead of a British Empire pith helmet, nurtured through the early decades of the twenty-first century. Mightn't it feel almost like turning back your personal clock fifty years? And mightn't it be wonderful to advise your junior copy on where you went wrong, and how to do it better? Isn't this, in (sometimes sadly) watered-down form, one of the motives that drives people to breed children in the ordinary way, by sexual reproduction?

If I have succeeded in my aim, you may be feeling warmer towards the idea of human cloning than before. But now think about the following. Who is most likely to get themselves cloned? A nice person like you? Or someone with power and influence like Saddam Hussein? A hero we'd all like to see more of, like David Attenborough? Or someone who can pay, like Rupert Murdoch? Worse, the technology might not be limited to single copies of the cloned individual. The imagination presents the all-too plausible spectre of *multiple* clones, regiments of identical individuals marching by the thousand, in lockstep to a Brave New Millennium. Phalanxes of identical little Hitlers goose-stepping to the same genetic drum—here is a vision so horrifying as to overshadow any lingering curiosity we might have over the final solution to the "nature or nurture" problem (for multiple cloning, to switch to the positive again, would certainly provide an elegant approach to that ancient conundrum). Science can open our eyes in both directions, towards negative as well as positive possibilities. It cannot tell us which way to turn, but it can help us to see what lies along the alternative paths.

Human cloning already happens by accident—not particularly often but often enough that we all know examples. Identical twins are true clones of each other, with the same genes. Hell's foundations don't quiver every time a pair of identical twins is born. Nobody has ever suggested that identical twins are zombies without individuality or personality. Of those who think anybody has a soul, none has ever suggested that identical twins lack one. So, the new discoveries announced from Edinburgh can't be *all* that radical in their moral and ethical implications.

Nevertheless, the possibility that adult humans might be cloned as babies has potential implications that society would do well to ponder before the reality catches up with us. Even if we could find a legal way of limiting the privilege to universally admired paragons, wouldn't a new Einstein, say, suffer terrible psychological problems? Wouldn't he be teased at school, tormented by unreasonable expectations of genius? But he might turn out even better than the paragon. Old Einstein, however outstanding his genes, had an ordinary education and had to waste his time earning a living in the patent office. Young Einstein could be given an education to match his genes and an inside track to make the best use of his talents from the start.

Turning back to the objections, wouldn't the first cloned child feel a bit of a freak? It would have a birth mother who was no relation, an identical brother or sister who might be old enough to be a great grandparent, and genetic parents perhaps long dead. On the other hand, the stigma of uniqueness is not a new problem, and it is not beyond our wit to solve it. Something like it arose for the first in vitro fertilized babies, yet now they are no longer called "test tube babies" and we hardly know who is one and who is not.

Cloning is said to be unnatural. It is of more academic than ethical interest, but there is a sense in which, to an evolutionary biologist, cloning is more natural than the sexual alternative. I speak of the famous paradox of sex, often called the twofold cost of sex, the cost of meiosis, or the cost of producing sons. I'll explain this, but briefly because it is quite well known. The selfish gene theorem, which treats an animal as a machine programmed to maximize the survival of copies of its genes, has become a favored way of expressing modern Darwinism (see, for example, Mark Ridley, 1996; Matt Ridley, 1996). The rationale, in one tautological sentence, is that all animals are descended from an unbroken line of ancestors who succeeded in passing on those very genes. From this point of view, at least when naively interpreted, sex is paradoxical because a mutant female who spontaneously switched to clonal reproduction would immediately be twice as successful as her sexual rivals. She would produce female offspring, each of whom would bear all her genes, not just half of them. Her grandchildren and more remote descendants, too, would be females containing 100 percent of her genes rather than one quarter, one eighth, and so on.

Our hypothetical mutant must be female rather than male, for an interesting reason which fundamentally amounts to economics. We assume that the number of offspring reared is limited by the economic resources poured into them, and that two nurturing parents can therefore rear twice as many as one single parent. The option of going it alone without a sexual partner is not open to males because single males are not geared up to bear the economic costs of rearing a child. This is especially clear in mammals where males lack a uterus and mammary glands. Even at the level of gametes, and over the whole animal kingdom, there is a basic economic imbalance between large, nutritious eggs and small, swimming sperm. A sperm is well equipped to find an egg. It is not economically equipped to grow on its own. Unlike an egg, it does not have the option of dispensing with the other gamete.

The economic imbalance between the sexes can be redressed

later in development, through the medium of paternal care. Many bird species are monogamous, with the male playing an approximately equal role in protecting and feeding the young. In such species the twofold cost of sex is at least substantially reduced. The hypothetical cloning female still exports her genes twice as efficiently to each child. But she has half as many children as her sexual rival, who benefits from the equal economic assistance of a male. The actual magnitude of the cost of sex will vary between twofold (where there is no paternal care) to zero (where the economic contribution of the father equals that of the mother, and the productivity in offspring of a couple is twice that of a single mother).

In most mammals paternal care is either nonexistent or too small to make much of a dent in the twofold cost of sex. Accordingly, from a Darwinian point of view, sex remains something of a paradox. It is, in a way, more "unnatural" than cloning. This piece of reasoning has been the starting point for an extensive theoretical literature with the more or less explicitly desperate aim of finding a benefit of sex sufficiently great to outweigh the twofold cost. A succession of books has tried, with no conspicuous success, to solve this riddle (Williams, 1975; Maynard Smith, 1978; Bell, 1982; Michod & Levin, 1988; Ridley, 1993). The consensus has not moved greatly in the twenty years since Williams's 1975 publication, which began:

> This book is written from a conviction that the prevalence of sexual reproduction in higher plants and animals is inconsistent with current evolutionary theory . . . there is a kind of crisis at hand in evolutionary biology. . . .

and ended:

> I am sure that many readers have already concluded that I really do not understand the role of sex in either organic or biotic

evolution. At least I can claim, on the basis of the conflicting views in the recent literature, the consolation of abundant company.

Nevertheless, outside the laboratory, asexual reproduction in mammals, as opposed to some lizards, fish, and various groups of invertebrates, has never been observed. It is quite possible that our ancestral lineage has not reproduced asexually for more than a billion years. There are good reasons for doubting that adult mammals will ever spontaneously clone themselves without artificial aid (Maynard Smith, 1988). So far removed from nature are the ingenious techniques of Dr. Wilmut and his colleagues; they can even make clones of *males* (by borrowing an ovum from a female and removing her own DNA from it). In the circumstances, notwithstanding Darwinian reasoning, ethicists might reasonably feel entitled to call human cloning unnatural.

I think we must beware of a reflex and unthinking antipathy, or "yuk reaction" to everything "unnatural." Certainly cloning is unprecedented among mammals, and certainly if it were widely adopted it would interfere with the natural course of the evolutionary process. But we've been interfering with human evolution ever since we set up social and economic machinery to support individuals who could not otherwise afford to reproduce, and most people don't regard that as self-evidently bad, although it is surely unnatural. It is unnatural to read books, or travel faster than we can run, or scuba dive. As the old joke says, "If God had intended us to fly, he'd never have given us the railway." It's unnatural to wear clothes, yet the people most likely to be scandalized at the unnaturalness of human cloning may be the very people most outraged by (natural) nudity. For good or ill, human cloning would have an impact on society, but it is not clear that it would be any more momentous than the introduction of antibiotics, vaccination, or efficient agriculture, or than the abolition of slavery.

If I am asked for a positive argument in favor of human cloning, my immediate response is to question where the onus of proof lies. There are general arguments based on individual liberty against prohibiting anything that people want to do, unless there is good reason why they should not. Sometimes, when it is hard to peer into the future and see the consequences of doing something new, there is an argument from simple prudence in favor of doing nothing, at least until we know more. If such an argument had been deployed against X rays, whose dangers were appreciated later than their benefits, a number of deaths from radiation sickness might have been averted. But we'd also be deprived of one of medicine's most lifesaving diagnostic tools.

Very often there are excellent reasons for opposing the "individual freedom" argument that people should be allowed to do whatever they want. A libertarian argument in favor of allowing people to play amplified music without restriction is easily countered on grounds of the nuisance and displeasure caused to others. Assuming that some people want to be cloned, the onus is on objectors to produce arguments to the effect that cloning would harm somebody, or some sentient being, or society or the planet at large. We have already seen some such arguments, for instance, that the young clone might feel embarrassed or overburdened by expectations. Notice that such arguments on behalf of the young clone must, in order to work, attribute to the young clone the sentiment, "I wish I had never been born because . . ." Such statements can be made, but they are hard to maintain, and the kind of people most likely to object to cloning are the very people least likely to favor the "I wish I didn't exist" style of argument when it is used in the abortion or the euthanasia debates. As for the harm that cloning might do to third parties, or to society at large, no doubt arguments can be mounted. But they must be strong enough to counter the general "freedom of the individual" presumption in favor of cloning. My suspicion is that it will prove hard to make the case that cloning does more harm to third parties than pop festivals, advertising hoardings, or mobile telephones in trains—to name three pet hates of my own. The fact that I hate something is not, in itself, sufficient justification for stopping others who wish to enjoy it. The onus is on the objectors to press a better objection. Personal prejudice, without supporting justification, is not enough.

A convention has grown up that prejudices based upon religion, as opposed to purely personal prejudices, are especially privileged, self-evidently exempt from the need for supporting argument. This is relevant to the present discussion, as I suspect that reflex antipathy to advances in reproductive technology is frequently, at bottom, religiously inspired. Of course people are entitled to their religious, or any other, convictions. But society should beware of assuming that when a conviction is religious this somehow entitles it to a special kind of respect, over and above the respect we should accord to personal prejudice of any other kind. This was brought home to me by media responses to Dolly.

A news story like Dolly's is always followed by a flurry of energetic press activity. Newspaper columnists sound off, solemnly or facetiously, occasionally intelligently. Radio and television producers seize the telephone and round up panels to discuss and debate the moral and legal issues. Some of these panelists are experts on the science, as you would expect and as is right and proper. Others are distinguished scholars of moral or legal philosophy, which is equally appropriate. Both these categories of person have been invited to the studio in their own right, because of their specialized knowledge or their proven ability to think intelligently and express themselves clearly. The arguments that they have with each other are usually illuminating and rewarding.

But there is another category of obligatory guest. There is the inevitable "representative" of the so-and-so "community," and of course we mustn't forget the "voice" from the such-and-such "tradition." Not to mince words, the religious lobby. Lobbies in the plural, I should say, because all the religions (or "cultures" as we are nowadays asked to call them) have their point of view, and they all have to be represented lest their respective "communities" feel slighted. This has the incidental effect of multiplying the sheer number of people in the studio, with consequent consumption, if not waste, of time. It also, I believe, often has the effect of lowering the level of expertise and intelligence in the studio. This is only to be expected, given that these spokesmen are chosen not because of their own qualifications in the field, or because they can think, but simply because they represent a particular section of the community.

Out of good manners I shall not mention names, but during the admirable Dolly's week of fame I took part in broadcast or televised discussions of cloning with several prominent religious leaders, and it was not edifying. One of the most eminent of these spokesmen, recently elevated to the House of Lords, got off to a flying start by refusing to shake hands with the women in the television studio, apparently for fear they might be menstruating or otherwise "unclean." They took the insult more graciously than I would have, and with the "respect" always bestowed on religious prejudice—but no other kind of prejudice. When the panel discussion got going, the woman in the chair, treating this bearded patriarch with great deference, asked him to spell out the harm that cloning might do, and he answered that atomic bombs were harmful. Yes indeed, no possibility of disagreement there. But wasn't the discussion supposed to be about cloning?

Since it was his choice to shift the discussion to atomic bombs, perhaps he knew more about physics than about biology? But no, having delivered himself of the daring falsehood that Einstein split the atom, the sage switched with confidence to geological history. He made the telling point that, since God labored six days and then rested on the seventh, scientists too ought to know when to call a halt. Now, either he really believed that the world was made in six days, in which case his ignorance alone disqualifies him from being taken seriously, or, as the chairwoman charitably suggested, he intended the point purely as an allegory—in which case it was a lousy allegory. Sometimes in life it is a good idea to stop, sometimes it is a good idea to go on. The trick is to decide *when* to stop. The allegory of God resting on the seventh day cannot, in itself, tell us whether we have reached the right point to stop in some particular case. As allegory, the six-day creation story is empty. As history, it is false. So why bring it up?

The representative of a rival religion on the same panel was frankly confused. He voiced the common fear that a human clone would lack individuality. It would not be a whole, separate human being but a mere soulless automaton. When I warned him that his words might be offensive to identical twins, he said that identical twins were a quite different case. Why? Because they occur naturally, rather than under artificial conditions. Once again, no disagreement about that. But weren't we talking about "individuality," and whether clones are "whole human beings" or soulless automata? How does the "naturalness" of their birth bear upon that question?

This religious spokesman seemed simply unable to grasp that there were two separate arguments going on: first, whether clones are autonomous individuals (in which case the analogy with identical twins is inescapable and his fear groundless); and second, whether there is something objectionable about artificial interference in the natural processes of reproduction (in which case other arguments should be deployed—and could have been—but weren't). I don't want to sound uncharitable, but I respectfully submit to the producers who put together these panels that merely being a spokesman for a particular "tradition," "culture" or "community" may not be enough. Isn't a certain minimal qualification in the IQ department desirable too?

On a different panel, this time for radio, yet another religious leader was similarly perplexed by identical twins. He too had "theological" grounds for fearing that a clone would not be a separate individual and would therefore lack "dignity." He was swiftly informed of the undisputed scientific fact that identical twins are clones of each other with the same genes, like Dolly except that Dolly's clone is older. Did he really mean to say that identical twins (and we all know some) lack the dignity of separate individuality? His reason for denying the relevance of the twin analogy was even odder than the previous one. Indeed it was transparently self-contradictory. He had great faith, he informed us, in the power of nurture over nature. Nurture is why identical twins are really different individuals. When you get to know a pair of twins, he concluded triumphantly, they even *look* a bit different.

Er, quite so. And if a pair of clones were separated by fifty years, wouldn't their respective nurtures be even *more* different? Haven't you just shot yourself in your theological foot? He just didn't get it—but after all he hadn't been chosen for his ability to follow an argument.

Religious lobbies, spokesmen of "traditions" and "communities," enjoy privileged access not only to the media but also to influential committees of the great and the good, to governments and school boards. Their views are regularly sought, and heard with exaggerated "respect," by parliamentary committees. You can be sure that, if a royal commission were set up to advise on cloning policy, religious lobbies would be prominently represented. Religious spokesmen and spokeswomen enjoy an inside track to influence and power which others have to earn through their own ability or expertise. What is the justification for this? Maybe there is a good reason, and I'm ready to be persuaded by it. But I find it hard to imagine what it could be.

To put it brutally and more generally, why has our society so meekly acquiesced in the idea that religious views have to be respected automatically and without question? If I want you to respect my views on politics, science or art, I have to earn that respect by argument, reason, eloquence, relevant knowledge. I have to withstand counterarguments from you. But if I have a view that is part of my religion, critics must respectfully tiptoe away or brave the indignation of society at large. Why are religious opinions off limits in this way? Why do we have to respect them, simply because they are religious?

It is also not clear how it is decided which of many mutually contradictory religions should be granted this unquestioned respect,
this unearned influence. If we decide to invite a Christian spokesman into the television studio or the royal commission, should it be
a Catholic or a Protestant, or do we have to have both to make it
fair? (In Northern Ireland the difference is, after all, important
enough to constitute a recognized motive for murder.) If we have
a Jew and a Muslim, must we have both Orthodox and Reformed,
both Shiite and Sunni? And then why not Moonies, Scientologists
and Druids?

Society accepts that parents have an automatic right to bring
their children up with particular religious opinions and can with-
draw them from, say, biology classes that teach evolution. Yet we'd
all be scandalized if children were withdrawn from art history
classes that teach about artists not to their parents' taste. We
meekly agree, if a student says, "Because of my religion I can't take
my final examination on the day appointed, so no matter what the
inconvenience, you'll have to set a special examination for me."
It is not obvious why we treat such a demand with any more respect than, say, "Because of my basketball match (or because of
my mother's birthday party, etc.) I can't take the examination on
a particular day." Such favored treatment for religious opinion
reaches its apogee in wartime. A highly intelligent and sincere individual who justifies his personal pacifism by deeply thought-out
moral philosophic arguments finds it hard to achieve conscientious
objector status. If only he had been born into a religion whose
scriptures forbid fighting, he'd have needed no other arguments at
all. It is the same unquestioned respect for religious leaders that
causes society to beat a path to their door whenever an issue like
cloning is in the air. Perhaps, instead, we should listen to those
whose words themselves justify our heeding them.

Science, to repeat, cannot tell us what is right or wrong. You cannot find rules for living the good life, or rules for the good governance of society, written in the book of nature. But it doesn't follow

from this that any other book, or any other discipline, can serve instead. There is a fallacious tendency to think that, because science cannot answer a particular kind of question, religion can. Where morals and values are concerned, there are no certain answers to be found in books. We have to grow up, decide what kind of society we want to live in and think through the difficult pragmatic problems of achieving it. If we have decided that a democratic, free society is what we want, it seems to follow that people's wishes should be obstructed only with good reason. In the case of human cloning, if some people want to do it, the onus is on those who would ban it to spell out what harm it would do, and to whom.

References

Bell, G. (1982). *The Masterpiece of Nature.* London: Croom Helm.

Dawkins, R. (1998). "The Values of Science and the Science of Values." In J. Ree and C. W. C. Williams (eds.), *The Values of Science: The Oxford Amnesty Lectures 1997.* New York: Westview.

Glover, J. (1984). *What Sort of People Should There Be?* London: Pelican.

Maynard Smith, J. (1978). *The Evolution of Sex.* Cambridge: Cambridge University Press.

Maynard Smith, J. (1988). "Why Sex?" In J. Maynard Smith (ed.), *Did Darwin Get It Right?* London: Penguin.

Michod, R. E., and Levin, B. R. (1988). *The Evolution of Sex.* Sunderland, Mass.: Sinauer.

Ridley, Mark (1996). *Evolution* (second edition). Oxford: Blackwell Scientific Publications.

Ridley, Matt (1993). *The Red Queen.* London: Viking.

Ridley, Matt (1996). *The Origins of Virtue.* London: Viking.

Williams, G. C. (1975). *Sex and Evolution.* Princeton, N.J.: Princeton University Press.

This article contains elements from shorter pieces that appeared in two London newspapers early in 1997, the *Evening Standard* and the *Independent.*

Soul Searching

George Johnson

Explorers returning from distant lands tell of aborigines so afraid of cameras that they recoil from the sight of a lens as if they were looking down the barrel of a gun. Taking their picture, they fear, is the same as stealing their soul.

You might as well just shoot them dead on the spot. Knowing that a photograph is only skin deep, people in the developed lands find such terror absurd. But the fear that one's very identity might be stolen, that one could cease to be an individual, runs deep even in places where cameras seem benign.

The queasiness many people feel over the news that a scientist in Scotland has made a carbon copy of a sheep comes down to this: If a cell can be taken from a human being and used to create a genetically identical double, then any of us could lose our uniqueness. One would no longer be a self.

There are plenty of other reasons to worry about this new divide the biologists have trampled across. Nightmare of the week goes to those who imagine docile flocks of enslaved clones raised for body parts. But the most fundamental fear is that the soul will be taken by this penetrating new photography called cloning. And here, at least, the notion is just as superstitious as the aborigines'. There is one part of life biotechnology will never touch. While it is possible to clone a body, it is impossible to clone a brain.

That each creature from microbe to man is unique in all the

world is amazing when you consider that every life-form is assembled from the same identical building blocks. Every electron in the universe is indistinguishable, by definition. You can't tell one from the other by examining it for nicks and scratches. All protons and all neutrons are also precisely the same.

And when you put these three kinds of particles together to make atoms, there is still no individuality. Every carbon atom and every hydrogen atom is identical. When atoms are strung together into complex molecules—the enzymes and other proteins—this uniformity begins to break down. Minor variations occur. But it is only at the next step up the ladder that something strange and wonderful happens. There are so many ways molecules can be combined into the complex little machines called cells that no two of them can be exactly alike.

Even cloned cells, with identical sets of genes, vary somewhat in shape or coloration. The variations are so subtle they can usually be ignored. But when cells are combined to form organisms, the differences become overwhelming. A threshold is crossed and individuality is born.

Two genetically identical twins inside a womb will unfold in slightly different ways. The shape of the kidneys or the curve of the skull won't be quite the same. The differences are small enough that an organ from one twin can probably be transplanted into the other. But with the organs called brains the differences become profound.

All a body's tissues—bone, skin, muscle, and so forth—are made by taking the same kind of cell and repeating it over and over again. But with brain tissue there is no such monotony. The precise layout of the cells, which neuron is connected to which, makes all the difference. Linked one with the other, through the junctions called synapses, neurons form the whorls of circuitry whose twists and turns make us who we are.

In the reigning metaphor, the genome, the coils of DNA that

carry the genetic information, can be thought of as a computer directing the assembly of the embryo. Back-of-the-envelope calculations show how much information a human genome contains and how much information is required to specify the trillions of connections in a single brain. The conclusion is inescapable: The problem of wiring up a brain is so complex that it is beyond the power of the genomic computer.

The best the genes can do is indicate the rough layout of the wiring, the general shape of the brain. Neurons, in this early stage, are thrown together more or less at random and then left to their own devices. After birth, experience makes and breaks connections, pruning the thicket into precise circuitry. From the very beginning, what's in the genes is different from what's in the brain. And the gulf continues to widen as the brain matures.

The genes still exert their influence—some of the brain's circuitry is hardwired from the start and immutable. People don't have to learn to want food or sex. But as the new connections form, the mind floating higher and higher above the genetic machinery like a helium balloon, people learn to circumvent the baser instincts in individual ways.

Even genetically identical twins, natural clones, are born with different neural tangles. Subtle variations in the way the connections were originally slapped together might make one twin particularly fascinated by twinkling lights, the other drawn to certain patterns of sounds. Even if the twins were kept in the same room for days, these natural predilections would drive them each in different directions. Experience, pouring in through the senses, would cause unique circuitry to form. Once the twins left the room, the differences between them would increase.

Send one twin around the block clockwise and the other counterclockwise and they would return with more divergent brains. For artificial clones the variations would accumulate even faster, for they would be born years apart, into different worlds.

Photography is only skin deep. Cloning is only gene deep. But what about the ultimate cloning—copying synapse by synapse a human brain?

If such a technological feat were ever possible, for one brief instant we might have two identical minds. But then suppose neuron No. 20478288 were to fire randomly in brain 1 and not in brain 2. The tiny spasm would set off a cascade that reshaped some circuitry, and there would be two individuals again.

We each carry in our heads complexity beyond imagining and beyond duplication. Even a hard-core materialist might agree that, in that sense, everyone has a soul.

This essay first appeared in The Week in Review of *The New York Times,* March 2, 1997.

PART II
Commentary

Sasha

Andrea Dworkin

would like to have my cat Sasha back again. He died maybe ten years ago. His death was terrible—episodic and prolonged. When he was taken to New York City's Animal Medical Center for an intractable case of constipation, the doctors admitted him to the hospital. A few days later he was deemed cured and John Stoltenberg and I could bring him home. But he didn't come home. Before releasing him, some attendants gave him a cold bath. He went into shock, and for days he lay nearly dead, tiny as if his body was sinking into itself and he was disappearing, smaller and smaller, so that eventually he would be soft bone and empty skin. He was so fragile that it hurt to look at him. John went to see him every day—from Brooklyn to the Upper East Side of Manhattan—and sat by the incubator in which he lay, talked to him, trying both to comfort him and to reach his will to live.

I think Sasha was there two weeks. He turned the corner to life, back to us, and one day John was able to bring him home. Our other cats—George, older, brilliant and dominating, and Cady, younger, female, quiet in her feline hauteur—were mean to him, because he smelled alien and different; and so Sasha sat alone on my bed, slowly recovering some strength and his native temperament, which was a beautiful and elegant sweetness, a tenderness of spirit. Sasha was kind. In cats or humans it is an uncommon quality. He had a big soul.

He never fully recovered. He was somehow weaker, a shade paler, less vibrant; and after not too long a time he had heart failure. Though we desperately tried to save him, he eventually could not sustain the effort of moving or breathing. We put him to sleep.

When I heard about the Scottish cloned sheep, I wondered if it meant that I could have had Sasha again, another Sasha but the same, my darling kind cat.

I have no warm spot for reproductive technologies of any kind except for contraception, condoms, and spermaticides. One mean theme of history, as I see it, is the ambition of men to control the reproductive capacities of women. Control of the whole woman was the best way to segregate her from a world of choice and possibility; and so in society after society, from the ancient Greeks to contemporary Saudis, women were and are imprisoned in homes and owned as reproductive chattel. While it continues to astonish me that men were and are willing to destroy the creativity and freedom of women in order to have children with whom they have, for the most part, little to do, I have no doubt that reproductive technologies continue rather than change this story of domination.

In *Right-wing Women* (Coward-McCann, 1983) I wrote about men's control of women's reproduction and sexuality in terms of what I called the *brothel model* and the *farming model:*

The brothel model relates to prostitution, narrowly defined; women collected together for the purposes of sex with men; women whose function is explicitly nonreproductive, almost antireproductive; sex animals in heat or pretending, showing themselves for sex, prancing around or posed for sex.

The farming model relates to motherhood, women as a class planted with the male seed and harvested; women used for the fruit they bear, like trees; women who run the gamut from prized cows to mangy dogs, from highbred horses to sad beasts of burden. (p. 174)

The farming model is extravagantly wasteful: Most women had to have children repeatedly so that some women would have some children that some men would want. In the brothel model, women are used for sexual release and pleasure—and their children are refuse, social garbage. The contemporary family headed by a single mother is stigmatized and disenfranchised because she has the social status of a whore, not a tamed, domesticated (married and fertile) female. (Whores are also pretty tame—do sexual tricks by rote—but men like to believe that in using whores they are walking on the wild side. And whores rarely have homes, even on sufferance, whereas farmed women are supposed to have homes—it's part of the deal.)

With the growth and legitimacy of the pornography industry in the United States and other Western countries, there is an increasingly potent social imperative that all women appropriate the sexuality of whores, which is to say, that all women be available and accessible to all men all the time and that all women submit (with a smile, unless gagged) to all brutal and exploitative sex acts, in private and in public, in "representation" and in life.

Of course, all women do not. Nonetheless, it is shocking to see a new desperation in women to be married mothers—to be in the most protected social circle—so as not to be exiled into the desolate dystopia of the pornographically liberated woman. Married, infertile women turn themselves into lab rats to achieve a technologized pregnancy. The process is painful and humiliating by all accounts; there is failure after failure; the woman's desperation to conceive grows with the physical torment of being worked on; and her sense of needing a baby more than she needs her own life can push her past decency, into the wombs of other women, less valued, women who will do (the new) reproductive prostitution because they need the money, the food, the shelter, all of which will be provided for the duration of the pregnancy. A live birth means the end of care and attention, thus the fairly common extortionate refusal to give up the baby. The stand-in, of course, may also feel love. Her body has fed the fetus; she has shared her blood, experienced morning sickness and an assault of hormonal change and disruption. She has carried the fetus inside herself, in her body, and then she is supposed to move on, childless from that pregnancy. This is cruel. This is wrong.

In a world in which cloning works, only compliant women will live. Cloning is the absolute power over reproduction that men have wanted and have destroyed generations upon generations of women to approximate. This, of course, is not the logical social consequence. The technology used to make the cloned sheep is perfectly adequate to induce parthenogenesis such that women could, if we choose, reproduce ourselves—and eventually this would be an all-female world, which would, probably, end at least rape, prostitution, incest, and forced pregnancy. Men would not have to be killed—an important point, since we seem so reluctant to kill them. They would just die out over time.

But they won't, will they? If they did not already have the real power over reproductive technologies, they would take it—using the violence that we will not use. But they do have it, don't they? They have it and they will use it. Women with attitudes will die or be killed or be exiled or marginalized to eventual death—well, just like now, but as transition, a gynocidal devastation. Within reach is a world with fewer but better women. They can be used exclusively for reproduction; they can be genetically copied; they will be captives as women mostly have been; and the defiant will be rooted out, the ambitious purged, the rebellious destroyed. Every man will be able to have the girl he wants when and how he wants her; and women finally will be less than human, as low and objectlike as men have wanted us to be, without will or freedom or dignity.

So, in the meantime, would I be able to have my Sasha back, another Sasha, to console me for the coming loss? In thinking about Sasha I understood that death is necessary. It forces change. It forces movement. Death makes it impossible for any of us to stand still. We try to hold on, but those we love die. We try to keep our small lives the same, but someone we cherish dies and our small lives change. Death forces us, while we live, to keep seeking, to risk, to love, someone different, not the same. To have another Sasha would be the first step toward stasis, toward turning life into a museum, a set-piece, a site of archeology and misplaced time. No, I must not be able to have another Sasha. He made me happy; losing him made me sad. Life pushes on, relentless, unsentimental; and we need passion, courage, and will to endure. Love and tenderness give endurance meaning, but we do not get to keep them.

And so I think the men who will clone the compliant women will control them both reproductively and sexually; and, in the process, they will destroy all human meaning: The men will abandon change for absolute control, any chance of intimacy for absolute power. Through cloning especially, men will defeat death; and change, too, will die. Life will be power without love or freedom or grace.

Sheep, Joking, Cloning and the Uncanny

William Ian Miller

The sheep has its own genre of joke, the sheep joke. This charming little subset of the vast domain of jokes has been around for centuries; the earliest instance I know of dates from an Icelandic saga of the thirteenth century. Cows, horses, dogs and cats cannot make this claim. They do not have their own genre of joke. Humans, however, do. Of course, it is we who are the chief butt of the sheep joke, since its defining subtext is the human male's lack of particularity as to choice of sexual object. With not much tweaking, the misanthropy of the sheep joke can easily be given a misogynistic bent as when the choice of sheep is not thrust upon some lonesome shepherd in the uplands removed from his conspecifics in the valley below, but when a so-called normal male actually prefers the comfort of sheep to the embrace of available women. Needless to say, there exists a Web site devoted to such tasteless jokes and one of their equally tasteless subgenres: the Scottish joke in which sheep also figure prominently.

Sheep jokes are sex jokes and although it may be the case that humans are not the only species that can joke, we may well be the only species that feels compelled to joke about sex. Even Dolly's creators, after all, were making a not very good sex (and sheep) joke, and perhaps an inadvertent Scots joke, when they rather tastelessly named her after Dolly Parton. These sophisticated researchers will still be boys (and Scotsmen). Sex and sheep seemed to come together by some inevitable association of ideas. The boy in them could not ignore prurient possibility in the fact that Dolly was generated from a cell taken from mammary tissue, a ewe's breast, 'tis true, but no matter; the fetishization of mammary tissue is not, it seems, particularly heedful of which mammal it comes from. The same Lockean association of ideas inevitably suggested Dolly Parton whose breasts, one suspects, also involved some scientific intervention. So we come full circle: sheep, breasts, sex, Dolly and science. The joke has even more twists: For where Dolly P's plastic surgeon merely made a breast into a larger breast, these researchers made a breast into a woman, of a different species to be sure, but still, as one form of the sheep joke has it, an object of desire in the heart of Midlothian.

With Dolly, however, the sheep joke of ancient lineage met the clone joke of more recent vintage. Some scholars may claim ancient pedigree for clone jokes too, citing Plautine comedies in which identical twins play havoc with the usual human presumption that we are all supposed to look enough like ourselves not to be mistaken for another, but let that pass as too scholarly a quibble even to engage scholars. The Plautine comedies suggest, however, no differently than clone jokes do, that when we are doubled it is a matter for joking (or as we will soon discuss, a matter for horror and disgust).

The meeting of sheep joke and clone joke is my theme or, if not quite my theme, it will be my point of departure. Sheep jokes have also of late been the theme of editorials in standard science journals. There the tone is of impatient contemptuous annoyance with the "terrible sheep jokes" in *Nature*'s words, which, to the editors' minds, along with sensationalized sci-fi horror riffs on cloning, trivialize the seriousness of the scientific issues raised by cloning. But then it was not the hoi polloi that began the sheep jokes, was it? It was Ian Wilmut himself, not only by naming Dolly Dolly, but by choosing a sheep rather than some other mammal whose reputation among humans would not have allowed for the easy coupling of our concerns about perverse sexuality and plainly reproductive sexuality, our anxiety about duplication with our anxieties about reproduction. In some unconscious stroke of inspiration, Wilmut, like the child of nature we suppose Shakespeare to have been, knew that literary and imaginative possibility were enhanced by the coming together of the sheep joke and the clone joke.

There were also other poetic impulses unconsciously and inexorably making the sheep the most appropriate choice of animal. Is it not a virtually universal belief that sheep, if not clones already, wish they were, or act so as to give us that impression? They long ago decided that there was no payoff in being a unique sheep. Doesn't the richly significant notion of the black sheep prove that? Even a mother sheep refuses or is simply unable to recognize her own offspring if it doesn't at least have a good chance of looking just like her. But do we think it a sign of sheep's stupidity that they recognize their own only when their children approach identity with themselves? Or do we rather think sheep much wiser, at the cost, it is true, of being much unkinder, than those poor little birds who stupidly raise the offspring of cowbirds only to have the thuggish impostors they raised push their real chicks out of the nest?

If we think of sheep as aggressively evincing a will to clonedom, as desperately aspiring to what comes as second nature to an amoeba and *E. coli,* we as humans are ambivalent, more than ambivalent. We are in many ways unnerved by the uncanniness of cloning, no less than we are by sex itself. And the most certain sign of our uneasiness, our being unnerved, is the compulsion to joke about it: In this case, we feel, perhaps, that this joking in the face of a new possibility for mass-produced life is in fact joking in the face of death, death of the spirit and death of the (male) body.

Joking (along with laughter, it seems) bears some inevitable, maybe even necessary, connection to certain vaguely uneasy and unnerving states: to the horrific, to the disgusting, to the uncanny on one hand and to the sacred, the pure and pious on the other. Joking seems to serve protective functions, by relief and release, as well as other ways; but it often finds itself paying a kind of inadvertent homage to what it mocks. The intense presence of the urge to joke is as sure an indication as there is that we are approaching the dangerous, the sacred and the magical. Pious and grave talk about human dignity is so often untrustworthy (we suspect it as the style of the hypocrite), so unfelt, so by rote, so safe and predictable, that some feel it necessary to retreat to the joke to pay serious homage.*

The merging of the sheep joke with the clone joke, in an indirect way, reveals just how much cloning appalls us, unnerves us, disgusts, horrifies and revolts us, precisely because it engages our deepest concerns about personhood, identity, life and sex. And if horror and disgust are too strong (not for me), there is no doubt that the possibility of perfect doubling disconcerts us, and suggests we are in the presence of the uncanny, however loosely we may want to understand that term.

Admit that there is something about duplication that partakes of the uncanny that even decades of xeroxing, nearly a century of assembly lines, a century and a half of photography and five centuries of the printing press have not quite inured us to. When young, we learn to grow used to our own reflection, even become fascinated by it; but it inverts us after a fashion, exchanging right and left so that our good side is to the Other our bad, our dexter is sinister to the Other. But when older, quite older, we are more likely to loathe what we see in the mirror if we even bother to at-

---

*I am seriously tempted to believe that the adolescent sick joke, the shocking joking about deformity, famine, death and degradation, is as certain a sign of real fear and respect of the positive norms at stake as the usual forms of piety might be. (Admittedly, the mockery is completely parasitical on the continued vitality of the usual forms of piety.) In adolescence, when our disgust mechanisms are at their most sensitive, we may have already achieved our most sensitive engagement with moral sensation. The engagement, to be sure, is via the *via negativa,* but it is moral engagement, not lack of it, that drives the joking. Contrast the maturer style of adulthood in which we often utter the right views in the accepted way as either a kind of cowardice or a kind of slothfulness.

tend to it. It is not being inverted that gets to us then (we never were really aware of the inversion except by way of finding that photos deviated from reflections in some small unnerving way), but representation itself, a representation that reveals the subversion of health and bodily decorum, makes known to us hairs that obtrude, spots that appear. We start to avoid mirrors because we feel that what we see is not just a representation of decay, but also a reproduction, that is, an augmentation, a doubling, of it.

We can tolerate photos of ourselves that flatter our most self-indulgent posturings in front of the mirror or before our mind's eye, but more primitive people, new to the technology, often feel that the photo captured their spirit, or perhaps reduced theirs by half at least. To show how we moderns still believe a representation of us somehow captures us, try ripping up a picture of yourself or poking a pin through its eyes or doing the same to a picture of your children or parents. It may be we have been cured of our ancient voodooism with regard to two-dimensional representations, although the fact that Dorian Gray's picture still inspires some dread suggests otherwise. Imagine, however, burning yourself or a loved one in effigy, rounded into 3-D? I suspect that even when a picture is being destroyed out of homage to the representee, because it is not a flattering likeness, there still lingers a little twinge or some small wonder that no twinge occurred. Note too that we blame the nonflattering photo not because it lies, but because it tells one kind of truth we do not like. Just as well-mannered people avoid the truth, so must well-mannered photographs.

Identical twins, although largely denatured of their danger, still are more likely to discombobulate us than "normals" do. The duplication is uncanny, which uncanniness can be experienced as a small twinge of the heebie-jeebies, or a small mirth or as a small occasion for wonderment or ambivalence. Some cultures find identicals so monstrous that they kill one or both; some cultures find them so special as to place them at the core of their foundational myths. But in either case they grab our attention, for they are manifestly disruptions in the proper ordering of things. And then they inspire in us all kinds of fantasy of what their experience must be: Imagine the limited horizons of certain kinds of self-deceptions that identicals must face. They must suffer seeing themselves exactly as others see them. I might be able to self-deceive regarding the attractiveness of the image strutting in the mirror, but what if I agree with everyone else that my identical brother whom no one can tell apart from me seems homely as a mudpuppy? I might deceive myself into thinking that the subtle differences between him and me make me the attractive sib, but how can I maintain that view when everyone calls me my brother's name? However, should my identical be obviously good looking, then I can assume a confidence about my self-presentation with an ease that comes seldom to normals.

Cloning from adult cells will prevent this particular kind of twin anxiety by interposing a generation or more between the twins. From the perspective of the older donor twin, seeing oneself as one once was will not be much more disconcerting than old home movies (which often have their own tale of woe and humiliation to tell). But what of the young clone looking up, knowing exactly what age and decay, what physical possibility holds in store. This is even worse I presume for strategies of self-deception than the ones identical twins must construct to make themselves believe they are better looking than their manifestly homely sib.

Sheep jokes tell us that we are painfully ambivalent about, I would even say scared out of our wits by, sex. It is not just the sex act or the rituals of courtship that frequently make either cowards or fools of us, it is also that we are more than a little ambivalent about the reproductive aspect of the sex act. Sex and reproductive fecundity immerse us fully in the realm of the disgusting. Not just the obvious bodily secretions and their minglings, not just the odors or unfortunate placement of the genitals, but reproductive fecundity itself leaves us immersed in too much flesh, too much ooze, too much life soup for many of our tastes. Now imagine the horrors of cloning—reduplication, perfect reduplication. Sure, we get rid of the messiness of sex or just turn it into the recreational sideshow most of us have always wanted it to be except for the three or four moments we decided it was time for kids. Suddenly fecundity and the disgustful side effects of it overwhelm us. Amoebae, algae, pond scum, fetid rank swampy ur-life. We replace the yuckiness of sexual reproduction (yes, I know it also has its allure, but most of that is parasitical on the yuckiness), with the yuckiness of pure pond scum. The last thing we need is more fertility, more ways of making more of us.

Contrary to most people's image of cloning as a fleshless, futuristic, hypersterile technological mode of "reproduction," I see it as the urge for an even more fecund fetidness than fleshly indulgent sex ever offered us. Cloning is about making us pond scum, with all its disgustful associations with excess, surfeit and eternal reduplicative recurrence. (Do we start to hear now the faint strains of the scary music that I wish to score my account with?) And we have not even added to the horror of amoebic excess, of triumphant bacteria, the horror of doubling, which I touched on briefly with my discussion of identical twins, but which I would like to repair to again for some added brief suggestions.

Doubling is invariably a theme of horror. What is it about effigies, wax museum figures, just plain dolls that are not toylike enough, that makes them sources of fear, nervousness or fascination? Even God is nervous about it; he feared images of Himself and prohibited them. The story is usually told that He did this because the image would be a false representation and the veneration of it would be a regression to idol worship. Idol worship was not His concern; it could hardly have been if He were omnipotent. His real concern was that He too would be subject to soul capture, that in being represented, he would be duplicated, doubled, that in the end the faithful would find God indistinguishable from his representation, or, worse, but a pale reflection of it; the faithful, He knew, could clone him more easily than He could ever let on. So he played the game in a minor key: He became the first to mimic vertical generational cloning, by making His son Him. The problem, we see, in representation is not that it represents, but that it in fact may duplicate.

But there are other fears that are prompted by doubling, among which we may include the horror that subtends Dorian Gray and on a more comic note, the incredibly bizarre and eerie self-refashioning of Michael Jackson as Diana Ross. Consider that the usual paranoid vision of duplication is a world of utterly fungible beings with no variation whatsoever. Anyone who has lived in a suburb or attended a rally of one's favorite cause knows we have pretty much achieved that already. The danger that is frequently alleged is that with perfect duplication, with perfectly identical looks at least, we will collapse into a gray undifferentiated mass. What will occur, I would bet, is exactly the opposite. The smallest difference will become charged with the greatest and most magical value. We will fight for differentiation with a vengeance and find it whether it is there or not. Academics, for instance, who are seen as perfect clones to the legislator downstate or upstate, manage to construct an honor culture in which some are esteemed and others ignored or mocked, with the subtlest differentiations in status noted, struggled for and worried about. Doubling then will give rise to a culture that will be ranked, be hierarchical and possess a foliated differentiation of exquisite subtlety.

Carry this a step farther as Dostoyevsky does in *The Double.* Suppose that one's perfect double should appear one day, even bear your name, work in your office, compete for the same rewards in the same circles. Suppose too, as Dostoyevsky does, that your double, your genetic twin, is loathsome, not to the sight, but morally reprehensible. He is insolent and mean to those beneath him, servile to those above, bowing and scraping in total disregard of his dignity. And then suppose, too, that precisely this flattering unctuousness causes him to advance at your expense. In the typical Dostoyevskian nightmarish and paranoid world the problem is not that one's double duplicates, but that with duplication comes, unavoidably and necessarily, *duplicity*. The double is the cheat, the crook, the shifty conniver who manipulates identity, your identity, to his advantage. The double runs up debts that you must discharge. Personhood decays into impersonation, is, in fact, indistinguishable from it. Cloning runs the risk of making us all impostures.

Pascal made the observation long ago that, as an existential matter, we humans are relegated to the middle of things; that we risk annihilation from the infinitely large and infinitely small. The ancient Norse cultures thought so too, putting us in middle earth, half way between the heat and cold, the giants and the gods. All this is a windy way of getting at the proper balance between representation, reproduction and duplication. Here I will personalize my account somewhat by making a confession. I have four kids, all of whom look like my wife. Sure, two of them have the shape of my feet. Three have my build. But they do not look like me, not in the least. But strangely they do not look like each other either, so that somehow my wife has the magical power to be looked like yet preserve their individuation with a vengeance.

I am, it should by now be clear, disgusted, even revolted by the idea of cloning: not just the idea of cloning humans, but the idea of cloning sheep too. I am quite frankly disgusted by Dolly. But here's the rub. I must admit that I also resent that my kids don't look like me, even though it is clearly to their social advantage that they favor my wife. Is it my egoism, my narcissism? I think it is rather my keen sense of honor and shame. My genes are just a bunch of wimps. They can't win any fight with my wife's genes unless the stakes are low, as in the battle for the exact shape of our kids' second toe. But it is also manifestly the case that I do not want my kids to look like me so that my paternity is clear, or that I can take some unfathomably greater pride in my reproductivity, or that I can just get a kind of uncanny feeling by seeing some vague resemblance to myself. I do not want identity; I just want some vague resemblance of the kind kids usually have to their parents. All I want, in other

words, is to be right in the middle of things. We weren't meant to
be duplicated except by a sibling born of the same mother and father within a few hours of each other and even then at a fairly low
probability; but neither are we meant to be erased from the not too
small statistical probability that we will be *vaguely* represented, not
duplicated, via sexual reproduction. We must also discount the
likelihood by the chances of infertility, of infant mortality, or, as in
my case, of the likelihood of sexually reproducing with a partner
whose genes outmuscle yours at every turn.

We were not meant to be duplicated across generations, only
represented vaguely. I know, saying such a thing makes me out to
be something of a know-nothing, who gives ready credence to an
immutable human nature and who believes in some divinely or sa-
tanically ordered telos. But I mean no such thing. All I mean to say
is that there are certain large constraints on being human and we
have certain emotions that tell us when we are pressing against
those constraints in a dangerous way. This is part of the job that disgust, horror and the sense of the uncanny do; they tell us when we
are leaving the human for something else; either downward toward the material, mechanical and bestial, or upward toward the
realm of spirit or the world of pure hokum. Nature gave us just
about as much doubling as we can handle without getting too
spooked.

Sameness Is All

Adam Phillips

It seems somehow appropriate—whatever the scientific equivalent of poetic justice may be—that the first animal to be successfully cloned was a sheep. Sheep, after all, are not famous for their idiosyncracy, for the uniqueness of their characters. We had assumed that sheep were virtually clones of each other; and now we have also been reminded that they are inevitably—all but two of them—genetically different. Now that what was once a figure of speech has become a reality—now that cloning has become a practicable possibility—it may be timely to wonder why describing someone as a clone has never been a compliment. That is to say, is cloning the death or the apotheosis of individualism?

One of the characteristics of contemporary culture has been a longing for community, for a sufficient sense of sameness with others; and at the same time a suspicion of people's wish to believe themselves to be too similar to each other. Or, indeed, identical to—overly convinced by—the images they have of themselves. From our experience of small-scale cults and large-scale fascism we have become fearful when too many people seem to agree with each other—seem to be of the same mind about something—or claim to know who they really are. Democracies, in other words, have to be ambivalent about consensus. Too little and there is fragmentation; too much and there is a (spurious) homogeneity. But with the advent of cloning—when the same has, as it were, been literalized as the identical, when the identical, at long last, supposedly exists—a whole range of political and psychological vocabularies are stopped in their tracks. Cloning, among many other things, seems to be a final solution to the problem of otherness. And, of course, the end of any continuing need—at least in the mass production of animals—for two sexes in the task of reproduction. In one fell swoop, cloning is a cure for sexuality and difference.

From a psychoanalytic point of view, one of the individual's formative projects, from childhood onwards, is to find a cure for—or, less strictly speaking, some kind of solution to—exactly these two things: sexuality and difference, the sources of unbearable conflict. And it is this that makes children's fascination with, and interpretation of cloning—and the whole science of genetics—so interesting. Indeed children should be consulted about all the great scientific issues of the day because they usefully anthropomorphize these issues for us (for them it's all about bodies and the connections between them: about mummy and daddy and me and you). So from the child's point of view, what might the appeal of cloning be, and what might be the conscious and unconscious fears associated with it? Since there could be no general answer to these questions, even if cloning itself radically changes our sense of what generalization might entail, I want to use two brief clinical vignettes that are suggestive and perplexing rather than exemplary— suggestive of what these particular children are using their ideas about cloning to say and to solve; and perplexing if for no other reason than the fact that one new thing adults can do now that children can't, but can aspire to do, is clone. In other (old) words—that once again cloning ironizes, shall we say?—children have a new role model on their horizon.

An eight-year-old girl who was referred to me for school phobia—which began a year after her sister was born—told me in her second session that when she grew up she was "going to do clothing." I said, "Make clothes for people?" and she said, "No, no cloth-

ing . . . you know, when you make everyone wear the same uniform, like the headmistress does . . . we learned about it in biology."
I said, "If everyone wears the same uniform, no one's special." She
thought about this for a bit and then said, "Yes, no one's special but
everyone's safe." I was thinking then, though I couldn't find a way
of saying it, that if everyone was the same there would be no envy;
but she interrupted my thoughts by saying, "The teacher told us that
when you do clothing you don't need a mummy and daddy, you just
need a scientist. A man . . . it's like twins. All the babies are the
same." There was so much in all this that I couldn't choose which
bit to pick up; I could only apparently carry on with the conversation. I said, "If your sister was exactly the same as you, maybe you
could go to school," and she said, "Yes," with some relish, "I could
be at home and school at the same time . . . everything!"

In this little girl's strong misreading—to use Harold Bloom's
phrase—of cloning, there seemed to be several theories afoot.
Firstly, that what she called "clothing" was a uniformity imposed by
a powerful solo individual, either the female headmistress or the
male scientist. And that if specialness or uniqueness was what was
lost at the birth of a sibling, then "clothing" was the last punitive act
in the drama—a drama that in actuality begins with the parents'
sexuality. You lose your place in the family when your sister arrives,
then you begin to experience your school uniform as the ultimate
proof of your loss of individuality. She wanted to "do clothing" perhaps because then she would at least be the active agent, not the
passive victim; the appeal here is of a certain kind of omnipotence
that in and of itself differentiates the cloner from the cloned. The
headmistress has one unique talent—to make the children the
same—but not the same as her. In other words, the scientist who
clones acquires a paradoxical (and enviable) uniqueness (as though
the new law of the genetic jungle is clone or be cloned).

The theory I seemed to want to introduce is that because she and
her sister are different, there is a competition (for the mother) that
she loses. Cloning, in other words, is a cure for the terrors and de-

lights of competition. If the two children were identical, they would, by definition, be getting the same things. If the parents don't have to have sex—"you don't need a mummy and daddy"—then the child doesn't need to be preoccupied by the relationship between the parents; nor indeed need she then see the primary relationship—the "source" that babies apparently come from—as one between two people of similar status, and that involves difference. My patient keeps her (symbolized) parents apart, the female headmistress and the male scientist, both doing the similar thing, but separately. She imagines one person having a mysterious talent, rather than two people doing something to each other that they can't do by themselves. What is interestingly obscure or ambiguous in this child's account is whether, or in what sense, the cloner and the clones are looked after—"you don't need a mummy and a daddy," as though the unconscious fear might be that the cloner is a tyrannical parent, oblivious or hostile to difference and the individuality of need.

It is also possible, of course, that the child is wondering about whether I am going to clone her, remake her out of my words. What Freud called *transference*—the way we invent new people on the basis of our earliest relationships—was evidence for him that a kind of psychic cloning went on between people, that we unwittingly treat people as though they were the same as us (the same as ourselves or our parents). The analyst interprets to show the patient that the analyst is not the same as anyone in his past, that the patient's cloning of the analyst is his defense against the shock of the new. The analyst, in other words, contests the patient's strong wish to clone him (psychoanalysis, one could say, was a cure for cloning before cloning itself existed, the cure as precursor of the problem). Psychoanalysis calls the simulation of sameness narcissism, which it tends to treat as the saboteur of development; Narcissus wanted to be the same as himself, the same as the image of himself, a distinction he didn't have it in himself to make. From a psychoanalytic point of view, successful psychic cloning—making people, including ourselves, in our own image of them—is a denial of difference and dependence, and therefore a refusal of need. The art of self-cloning is an attempt to stop time by killing desire. Replicating myself, I keep finding nothing else. I depend on something other than myself to actually nourish me.

Adolescents, of course, are preoccupied by the relationship between dependence and conformity, between independence and compliance. So it's not surprising perhaps that in the cultic jargon of adolescents, to be called a clone is an ultimate insult. And yet it may be worth wondering exactly what is being repudiated—what longings or pleasures may be encoded in the word—in this particular, increasingly topical, form of scapegoating. In other words, cloning gives us the opportunity to rewonder what's wrong with being apparently or exactly the same as someone else. And this is a question not only about why we may be frightened of being like other people—what it is we imagine we lose in this process—but about why we may be frightened of other people being just like us.

If cloning as analogy captures the adolescent imagination, then this tells us as much about cloning as it does about adolescence. And one thing it tells us, I think, is that there is, for some people, a deep fear of not being a clone, of not being identical to someone, or identical to someone else's wishes for oneself (the child enacting the parents' conscious and unconscious projects being the model for this). As though we can only work out what or who we are like from the foundational belief—the unconscious assumption—that there is someone else that we are exactly like. For the adolescent the question is: If I'm not the same as someone else, what will I be like? And this is where cloning comes in. The adolescent lives as if she has been cloned, and is trying and not trying to find out what else there is to her. But now, of course, she can also find a more accurate and exacting description—an unprecedented picture—of this predicament which recreates it anew. Once there were twins and mirrors and doubles; now cloning is available for adolescent consumption. Ideas about development are themselves developed by fresh analogy.

A sixteen-year-old boy given to disparaging some of his male contemporaries as clones—kids who, in his view, were rather too keen to please their teachers, rather too timid to defy their parents—mentioned one day in a session that he wanted to find a girlfriend who would be his clone: "Just like me so we'd have lots of things in common . . . anyway, girls are so good at being clones." I asked him what that meant, that they were good at, and he replied, "Being like the people they like." I wondered aloud whether if this was true, it was because some girls picked up how frightened boys were of them. And he said, a bit too quickly, with that dazzling logic we all have recourse to when ruffled, "Well I'm not scared of them, that's why I want a clone girlfriend." I suggested that having things in common with people was overrated. This interested him, so after a pause I continued with this impromptu lecture that was stirring in me, elaborating what I meant—because he had asked—and ending, for some reason, with a slightly impatient question: "Anyway, what would two people who were exactly the same do together?" I was, at this moment, genuinely confounded by this and didn't expect him to answer (it's often surprisingly difficult to work out who one's questions are addressed to). We both sat there for a few minutes and then he said, as if he was stumbling over what I'd said, "They couldn't ask each other . . . because they'd get the same answer, so they'd have to ask someone else." And I said, clearly helped by him, "So they'd need a third person. They'd still need somebody different?" And he said, "Yes," perking up, "Me and my clone would need you to guide us!"

The fantasy of the clone girlfriend—not exactly a rarity—was for this boy an all-purpose magical solution, a way of preempting what you do about, or with, the parts of yourself that have nothing "in common" with an object of desire. What is of interest is that the (narcissistic) solution of creating absolute sameness, the clone, unconsciously kills desire. The fantasy of cloning a girlfriend is a fantasy of not needing a girlfriend. The exact replication of the self merely replicates the problem. That men must not be clones, but must clone women—one contemporary description of a war between the sexes—suggests, among other things, that men already experience themselves as having been cloned by women. The fantasy of cloning—made puzzlingly real by the actuality of it—raises the question now of what the alternative analogies are for relationship? The cloning has already happened; the problem is where we can go from here?

Cloning is, for obvious and not quite so obvious reasons, a compelling way of talking about what goes on between people. But of course the fantasies about cloning—the cultural gossip about it—are informed by wishes. Genetic inheritance, for example, is always a potential: It doesn't predict growth, but it provides (often unknowable) constraints. The environment is an essential part of the equation. The cloned sheep will be identical genetically, but they will have different histories. As my two examples suggest—though for quite different reasons given the ages of the children—cloning is used to get around history, as though in the total fantasy of cloning, history as difference is abolished. People, in actuality, can never be identical to each other. Perhaps this relentless wish for absolute identity—that even real cloning cannot satisfy—conceals, tries to talk us out of, a profound doubt about our being the same as anything. Wishes always return us to the scene of a crime.

Queer Clones

William N. Eskridge, Jr.
Edward Stein

Reproduction has been part of the controversies surrounding homosexuality for as long as there have been such controversies. Homophobes of various stripes (as well as others) have been concerned that condoning homosexuality might lead to a decline in the ability of the human species to reproduce, while at the same time being worried about the ways in which homosexuals might try to reproduce themselves. Lesbians, gay men, and bisexuals (collectively, queers),[1] for their part, often want to have children, typically for the same, other-regarding reasons other people desire them. Lesbian and gay parents have thereby exercised their right to create "families we choose," less encumbered by traditionalist notions of family, marriage, and the like. Queer families have, in turn, been a source of concern to some homophobes who are affronted that lesbians and gay men would appropriate heterosexual institutions, including marriage and childrearing.[2]

The children in queer families of choice sometimes were begat in previous relationships with a partner of the opposite sex. Increasingly, children are begat within a same-sex union through artificial insemination, various forms of surrogacy, co-parenting schemes, and adoption.[3] The law has not been friendly to such families; courts and agencies in various states regularly deprive gay and lesbian parents of custody of their children at the behest of nongay co-parents or even relatives, deny adoption rights to lesbian and gay individuals or couples, and create legal difficulties for gay sur-rogacy and lesbian artificial insemination efforts.[4]

Given the difficulties, many of them legal, with these various re-productive methods, the advent of a reliable technique for human cloning would, at first glance, offer queer families of choice a fas-cinating and useful alternative for reproduction. Under this tech-nology, already explicitly accomplished with sheep,[5] genetic material would be taken from one person (call her the *clonee*) and grafted into an egg of another person (call her the *ovum surrogate*) in such a way as to create the equivalent of a fertilized egg (a zy-gote). This zygote would have the same genetic makeup as the clonee. The zygote would contain none of the genetic material of the ovum surrogate, which would only provide the "structural" material, the "shell," for the zygote. The resulting zygote would then be implanted into the womb of yet a third person (call her the *gestational surrogate*) and would develop into a fetus and then be delivered as a human baby. There seems no biological reason why the roles of clonee, ovum surrogate, and gestational surrogate can-not be combined, and one female could provide the genetic mate-rial, the "shell" for the zygote, and the womb in which the zygote gestates.

Under cloning technology as we imagine it, a single lesbian or bisexual woman, or a gay or bisexual man or a transsexual[6] with the assistance of a woman, could be the clonee parent of a child with her or his genetic material. Furthermore, once the technology ex-ists for grafting genetic material into an egg to produce a zygote, there should follow technology for *gene-splicing,* whereby the genes of two people of the same sex could be spliced together to produce a zygote that shares genetic material with both. With the advent of gene-splicing, same-sex unions could not only produce children, but also produce children who are genetic hybrids of the parents, just like those produced in different-sex unions. Cloning and gene-splicing would also offer reproductive possibilities for transsexuals and people infected with the virus that causes AIDS. These persons often cannot easily reproduce or reproduce at all under current technologies; cloning would offer a better and safer (for infected people) means for them to have children, and gene-splicing would offer a broad array of partnerings fresh possibilities for reproduction.

The implications of these scientific and social developments are as unpredictable as they are profound. This essay explores some implications whose contours are suggested by reflection on the practical, legal, historical, and biological features of lesbian and gay families of choice. (Although our analysis, especially our legal analysis, will focus on the United States, some of the general points should have broader, transnational application.) These implications flow from the following ideas. To begin with, queer cloning and gene-splicing will almost certainly be easier (both mechanically and legally) for lesbians than for gay men. Thus, one result of the development of these technologies will be to reinvigorate the feminist utopian idea of women reproducing without men. Moreover, some queer people may be attracted to cloning because they think it will be a way to have queer children. The thought is that a child who is the product of queer cloning—because he or she would be genetically linked only to a queer parent (in the case of cloning) or two queer parents (in the case of a couple opting for gene-splicing)—would inevitably or very likely be queer as well.

Finally, and most importantly, cloning has the potential to complete the radical transformation that queer people offer society: Gay people have always engaged in *mutual sex without reproduction,* a pattern now typical of straight people as well; with cloning, gay people (and others) would have the option of engaging in *mutual reproduction without sex.* This would expand the options that lesbians and gay men have for reproducing. This complete separation would also have far-reaching implications for the diversity of families in America. For people infected with the virus that causes AIDS and for transsexuals, cloning and gene-splicing will in some cases allow people to reproduce who would not otherwise have that option.

Gendered Asymmetries

The biology of cloning and the current legal regimes ensure that *queer cloning,* by which we mean the use of cloning or gene-splicing technologies by lesbians, gay men, bisexuals, and transsexuals, will be much easier for lesbians than for gay men and for transsexuals. Assuming the development of the necessary technology, a lesbian who wants to clone herself could carry her own clone to term, and the current law in every American jurisdiction would recognize her as the mother of the child. There would, of course, be no legal father (unless the lesbian were married to a man). In an increasing number of jurisdictions, however, the lesbian's female partner (if she had one) could petition for a "second-parent adoption," in which she would legally become the child's second parent.[7] Current law does not, however, provide for gene-splicing. Under current law, if two women together provide the genetic material for a child and one of them carries the baby to term, the woman who plays the gestational role would be a legal mother, but it is not clear what legal rights the other genetic mother would have. The law in these cases should follow the policy of the second-parent adoption cases and recognize *both* women as legal as well as the biological mothers. Like the second-parent adoption decisions, however, this approach would require a creative interpretation of the governing statutes, a move that some judges will not be willing to take.

Queer cloning would be a boon to lesbians and would probably contribute to an even more pronounced lesbian baby boom.[8] Under current (that is, precloning) practices, many lesbians are reluctant to have children by known sperm donors because of legitimate fears that the male donors will assert parental rights and interfere with their raising of the children, or that the male donors may carry a disease (prominently, the virus that causes AIDS) that could be passed on through insemination. Women having these concerns can turn to sperm banks, but some sperm banks will not deal with unmarried women. Moreover, some women are uninterested in sperm banks, because they would prefer to know the person who is the genetic partner in creating their children. The advantage of cloning (or gene-splicing) is that sperm—and hence men—become unnecessary. With cloning, a lesbian with a functional womb could clone herself and carry the resulting zygote to term or, if she wants to mix her genes with those of her life partner or a close friend, gene-splicing could do the trick.

Queer cloning, therefore, makes possible the feminist utopian notion, propounded by some women in the early 1970s, of a community of women, unencumbered by men. To be sure, not all women shared this vision in the 1970s, and most probably do not today, but with the advent of queer cloning this possibility will become a live one. Technological change might broaden the appeal of women's communities, not just for lesbians and bisexual women, but also for many heterosexual and asexual women. Nonqueer women would also have the option of having children with other women, but without having sex with them. If this occurred, we wonder whether the concept of sexual orientation itself might not carry diminished power or even disappear over time.

The foregoing is at this point speculative, because queer cloning is neither technologically available for humans nor currently legal. Even if they were available in the foreseeable future, such reproductive technologies would be quite expensive, and health insurance companies would be unlikely to pay for use of such expensive technologies. In the short term, therefore, queer cloning would only be available to the wealthiest women. And, contrary to popular myth, lesbians (at least those who identify themselves as such) are at the bottom of the income scale, making much less income on average than straight or gay men and somewhat less than straight women.[9]

Whatever the limitations of cloning and gene-splicing for lesbians, they are magnified for gay men and for most transsexuals.

The main impediment is practical: Even under cloning technology as currently envisioned, gay men and male-to-female (and some female-to-male) transsexuals would still need the cooperation of a woman to bear a child. While cloning would be able to create a zygote from one male or, with gene-splicing, two males, the zygote would still need a human womb in which to gestate. In the foreseeable future, therefore, men would still need women to make use of cloning, even though women would not need men. This practical fact, alone, would discourage gay men and transsexuals from queer cloning in the foreseeable future, for they would have to find a woman willing to carry the embryo to term. In most cases, such an arrangement would involve a fee. Thus, all the expenses women would bear in such a cloning arrangement would be borne by those seeking surrogates, who would also bear the additional expenses of finding and compensating a woman to carry the child. Only well-to-do people would be able to afford this. There is also the possibility (although this is more speculative) that what might be called *incubation* technology will advance to the point whereby a human zygote could be gestated inside a machine, a female of another species, or even a human male. Even if this becomes possible and men are able to reproduce without women, the costs involved would be exceedingly high. While the myth of gay wealth holds in part for men—gay men's average income is higher than that of straight women or lesbians—gay men on average earn less than straight men.[10]

Setting aside the possibility of gestation without females (there are, of course, no legal structures for dealing with offspring who have no gestational mother and no female genetic parent), there is also a legal impediment to queer cloning for men and most transsexuals.[11] Surrogacy arrangements, where a person pays a woman to be impregnated and bear a child, are unenforceable in most jurisdictions, such as New York and New Jersey,[12] and criminally illegal in some, such as the District of Columbia and Michigan.[13] In other jurisdictions, such as Florida and Virginia, a surrogacy arrangement is only enforceable if the recipients of the child are a married couple, a requirement that excludes most gay men (as well as lesbians who want to make use of surrogacy because they are unable or unwilling to be pregnant).[14] In other jurisdictions, the state of the law is unclear or in flux. In California, Governor Wilson in 1993 vetoed a bill that would have explicitly allowed surrogacy arrangements. Does that mean that such arrangements are void? In the same year, the California Supreme Court held that gestational surrogacy is permissible under state law.[15] Does that reasoning apply to queer cloning? The law is not completely clear in California, and it is substantially less clear in the many other jurisdictions where there are no statutes or judicial decisions addressing surrogacy issues.

What is clear is that the regulatory schemes in place for virtually all American jurisdictions were conceived with an infertile straight couple in mind; male couples fall outside, or not clearly inside, their ambits. "Surro-gay" arrangements, as Marla Hollandsworth calls them, are carried out every week in the United States, but usually "around" rather than "within" the law. Without the force of law, such arrangements are unenforceable, which means that the surrogate mother can decide to keep the baby to which she has contributed genetic material. Would this presumptive policy be extended to gestational surrogates? Logically not, as California's Supreme Court held, but logic is not the surest guide in this area of law. At the very least, until the law is clarified, queer cloning by gay men runs the risk of losing the child to the gestational mother in most American jurisdictions. This lack of clarity increases the legal risk, and probably the price as well, of queer cloning and could be expected to discourage gay men and transsexuals from taking advantage of it.

Even in states that ultimately allow surrogacy arrangements, parentage statutes may create problems for gay men and transsexuals seeking to clone. When the surrogate is a married woman and the arrangement is carried out under a physician's guidance, as many as half the states provide by statute that the woman's husband, and not the sperm donor, is the legal father of the child.[16] This is an odd legal regime in the case of gay cloning (or any instance of gestational surrogacy), for neither "legal" parent would have any biological connection with the child, and the only person(s) with a biological connection, the gay man or the transsexual, would have no legal rights. This would not be an inevitable result even in those states, because it is likely in at least some of those states that (1) a gestational surrogate is not regulated, (2) the legal regime can be avoided by using unmarried women as surrogates, or (3) the legal regime can be avoided by not using a licensed physician, as the California courts have held.[17]

In addition to the practical and legal difficulties of queer cloning by gay men, there are moral difficulties. Some gay men, including the authors of this essay, are feminists and would be ethically concerned about critiques of surrogacy as exploiting women's bodies, commodifying an inalienable feature of female personhood, and reinforcing gender stereotypes of women as "breeders."[18] On the other hand, many feminists powerfully defend surrogacy as a freedom that women have to deploy their bodies; some of the same prochoice arguments that support the right to abortion also support the right to surrogacy.[19] The feminist debate itself presents intractable ethical issues (which we do not attempt to resolve here); other normative as well as descriptive theories further complicate matters.

The moral quandaries buttress our argument that queer cloning would be a less desirable option for gay men and most transsexuals than for lesbians. Further, queer cloning might undermine current options gay men have for forming families of choice. Under current practice, where sperm is needed for reproduction, gay men are able to form family clusters with lesbians; that is, one or two gay men contribute sperm to one or two lesbians, and the child is raised with gay and lesbian parents—often two mommies and two daddies. If it comes to pass that sperm is no longer needed to reproduce, men would no longer be biologically necessary, and the impetus for such family clusters would diminish. (It would not disappear if women believed that, in our culture, having a father as well as a mother is psychologically important to the child or if queer cloning was prohibitively expensive for many people.)

The Perpetuation of Queer Culture?

People want to have children for all sorts of reasons. The reasons that lesbians, gay men, and bisexuals have for wanting children are pretty much the same as the reasons why straight people want to have children—the joy of nurturing another human being, the perpetuation of the species, carrying on an extended family, and providing one's own parents with grandchildren. One distinctive reason some lesbians and gay men desire children involves their desire to perpetuate queer culture by having queer children. Such people usually operate under the assumption that gay people will have gay children or that they are more likely to do so than straight people. If this is true because sexual orientation has a genetic component, then queer cloning offers a way for queer people to produce queer children, or at least to increase the odds of doing so. If I am gay, surely my clone will also be gay, right? Not necessarily. Like so many of the other myths about sexual orientation, this one is beset by complexities, inaccurate beliefs, and unanswered questions.

A person and his or her clone are genetically identical, but that does not mean they will share all of the same traits. A person's physical and psychological characteristics are not simply read off a person's genes, not even on the most genetically deterministic view of things; genes in themselves do not directly specify any behavior or psychological phenomenon. Instead, genes direct patterns of RNA synthesis, that in turn specify the production of proteins, that in turn may influence the development of a psychological characteristic, that in turn causes behaviors. At every step of this process linking genes and psychological properties or behaviors, nongenetic factors (environmental factors, broadly construed) play a role. A person's development is significantly affected by a wide range of nongenetic factors, such as prenatal hormonal levels and diet during puberty, to pick two among many examples. The upshot with respect to sexual orientation is that, due to the panoply of developmental factors, it is quite possible for genetically identical individuals to have different sexual orientations.

Concrete evidence of this can be found by looking at twin studies of sexual orientation. The most recent and sophisticated twin study, done by Michael Bailey, the most prominent researcher using twins to study sexual orientation, reported that between 20 and 38 percent (depending on how broad a notion of being gay or bisexual is used) of the identical twins of gay and bisexual men (and who have identical twins) were also gay or bisexual and that between 24 and 30 percent of identical twins of gay and bisexual women were also gay or bisexual.[20] This is lower than the percentage reported in earlier studies by Bailey, which showed that about 50 percent of the identical twins of gay and bisexual people were also gay or bisexual.[21] (Bailey's own assessment of his earlier studies is that the population sample he used may have been biased.)[22] The most interesting and consistent finding of recent twin studies is that at least half of identical twins had different sexual orientations.[23] This is the case even though the identical twins in these studies shared all of their genes and most environmental factors. It is not clear how to use twin studies to disentangle the genetic and environmental influences that operate on identical twins.[24] Genetic factors no doubt play *some* role in shaping a person's sexual orientation (similarly, genetic factors play *some* role in shaping a person's dietary and religious preferences, for example), but how much of a role they play and how the role is mediated are far from clear. The consequence for queer cloning is that neither cloning nor gene-splicing of queer people will guarantee queer children. Given the role that the environment is likely to play in the development of sexual orientation and the fact that sexual orientation is a cognitively mediated property, it is not clear that queer cloning will produce queer children. We think there is not even enough evidence to speculate about the odds.

Nonetheless, queer cloning could improve the conditions of lesbians and gay men, perhaps substantially so. The biggest improvement could be that cloning would provide a way for some gay men and lesbians, who would otherwise not be able to have children, to do so. While it would be foolish for them to have children in order to replicate their sexual orientation, few queer people want children for this reason only, and most queer people with children find the nurturing, sharing, and other generative experiences to be among the most rewarding of their lives—just as straight people do. Moreover, even though queer cloning would not necessarily produce more queer children, there is good reason to think it will contribute to a more "queer-friendly" culture in general. Social scientific studies and anecdotal evidence suggest that the (non-clone) offspring of gay men and lesbians are more likely than people in general to be queer-friendly, that is, nonhomophobic, more supportive of lesbian and gay rights, and so forth.[25] Queer parenting, including cloning, would then have benefits for the queer community and would help in the development and maintenance of a just society.

Ideological Land Mines

Queer cloning offers the possibility of completing the perfect segregation of sex and reproduction that is already increasingly characteristic of lesbian and gay couples. Same-sex sexual activity does not lead to reproduction, as everybody already knows; with the advent of queer cloning, queer reproduction could be completely divorced from sex and even the mechanical simulation of sex, namely, artificial insemination. Queer cloning is also, potentially, the apotheosis of some versions of lesbian feminism. It would further fuel the lesbian baby boom, and maybe initiate a tiny transsexual baby boomlet. Finally, even skeptics such as the authors of this essay concede that there is *some* kind of genetic component to sexual orientation (but whether it is primarily or significantly genetic is unproved at this point, but not refuted either). Given this, queer cloning could be a way to increase the relative number of queers in our society.

Any potential benefits of queer cloning will inevitably be tempered by individual and societal homophobia. The depth of homophobia in American culture derives in part from the way in which it is cumulative of several related anxieties: In the minds or hearts of many Americans, same-sex intimacy violates natural law (whether defined religiously or philosophically) because it involves sex without reproduction, inverts proper gender roles by suggesting to women that they do not need men for their well-being, and is sexually threatening to straight people, whom "homosexuals" allegedly prey upon or recruit.[26] Americans, in general, tend to be nervous about the breakdown of traditional standards of sexual and gender morality *or* women's taking over men's traditional roles and leaving the home *or* the unleashing of the libido. This nervousness is likely to manifest itself as anxiety towards queer people— as outside traditional sexual morals, as inherently gender-bending (partly because many people connect homosexuality with sissy boys, butch girls, boys who want to be girls, and vice versa), and as more sexualized than other people. Nervousness with respect to all three phenomena has in the past produced hysterical manias and witch-hunts against queer people.

If queer cloning occurred on a significant enough scale that mainstream society attended to it, many would certainly be alarmed, because queer cloning would trigger all the anxieties that inspire homophobia in so many Americans. That same-sex couples not only have "unnatural sex" (that is, sex unlinked to reproduction) is bad enough to a traditionalist, but that they might also have "un-natural reproduction" (that is, reproduction unlinked to sex) is even worse. That lesbians can abandon relations with, and depen-dence on, straight men is bad enough to a sexist, but that they do not even need men (sperm) for the most male of activities (im-pregnation) is not only worse but a catastrophe. That homosexu-als not only allegedly threaten children with molestation and recruitment is bad enough to the homophobic parent, but that they can create their own recruits as well is, to such a person, a social nightmare. The worst, and most far-fetched, hetero-nightmare is that, after a few generations of self-procreating homosexuals, queers might outnumber straight people and subject straights to the same sorts of discrimination and abuse to which homosexuals have traditionally been subjected.

These comments are not abstract musings. This sort of thinking is currently propelling social and legal reaction. In 1993, Hawaii's Supreme Court held that the state prohibition of same-sex mar-riage is sex discrimination that must be justified by a neutral and strong state interest, a task even committed homophobes recognize as unlikely.[27] Even though American society allows divorce on de-mand, tolerates adultery and fornication, and does too little to ad-dress the violence of rape and molestation of women and girls, many people have become so enflamed at the mere possibility of same-sex marriage that one state legislature after another has en-acted measures to ensure that same-sex marriages will not be rec-ognized in their jurisdictions, and Congress in 1996 enacted the Defense of Marriage Act (DOMA) that not only reassured the states the power to avoid recognition, but also assured that 1049 federal statutes would never be construed to benefit the same-sex couples that Hawaii might join in matrimony.[28] In a nation where so many more tangible problems are pressing, the legislative re-sponse to the possibility of same-sex marriage has been astounding. It is comprehensible only in light of the fact that same-sex marriage triggers all the anxieties that contribute to homophobia: It would be unnatural; marriage "must" be gendered; "real" marriages would be threatened (and were "defended" in 1996 with the passage of the DOMA).

As with the *possibility* of same-sex marriage, the *possibility* of queer cloning might be a stimulus for a preemptive antigay reaction. States could, for example, criminalize or void gestational surrogacy unless the intended parents were a married couple. Such measures would only affect gay male or transsexual cloning; it is doubtful that state family law could much affect lesbian cloning, for if a state tried to do so, lesbian cloning would just go underground. To deal with lesbian cloning, the federal government could prohibit gene-splicing or cloning altogether or limit such methods to different-sex couples, although the latter is less likely because of constitutional objections. Under some circumstances, it is also conceivable that queer cloning would contribute to a more general antigay backlash. Conversely, antigay sentiments probably contribute to anticloning sentiments. Because cloning would, in fact, be the apotheosis of "unnatural" reproduction and would, in fact, free up women to reproduce without the aid of men and would, in some people's fantasies, allow narcissists to reproduce themselves, some of the same impulses that form intense homophobic reactions would generate reactions to the possibility of cloning: Thus, cloning is unnatural, it divides the sexes, people who do it are self-absorbed.

Another way that the advent of cloning and related technologies might positively harm queer people involves potential attempts to use such technologies to prevent the birth of queer or potentially queer children. Some heterosexuals might use cloning or gene-splicing to try to ensure that their children would be heterosexual. Given the widespread prejudice and discrimination against lesbians, gay men, and bisexuals, it seems quite likely that many parents will try to ensure that their children are heterosexual. This is especially plausible, as parents worldwide sometimes choose to abort a fetus if it is not of the sex they desire.[29] Using cloning technologies (as well as other genetic interventions) to prevent the birth of queer children treats lesbians, gay men, bisexuals, and transsexuals as diseased, not worthy of living, and the like. The advent of cloning could lead to a proliferation of such attitudes. Such proliferation would increase the pressure to hide one's homosexuality and would decrease the collective power of lesbians and gay men. The availability of cloning and splicing procedures that some might use to try to ensure that their children would be heterosexual would engender and perpetuate (1) attitudes that lesbians and gay men are undesirable and not valuable, (2) policies that discriminate against lesbians and gay men, and (3) the very conditions that give rise to such attitudes and policies. Even if such procedures do not work, which they would not (for the same reasons that queer cloning would not guarantee queer children), their mere availability may well engender these results. Biological procedures do not have to be valid to be implemented.[30]

We doubt that any state regulatory response would be the end of queer cloning (or cloning to ensure heterosexuality), unless a worldwide effort simply halted the development of cloning and gene-splicing technologies. Once the technology becomes available anywhere in the Western world, it will become available to some Americans. Thus, lesbians with sufficient resources could travel to France or traffic in domestic black-market cloning or gene-splicing to create their families if they really desire; some gay men could try some of these approaches but with greater trouble and expense.

*   *   *

Queer cloning can be viewed as the next logical step in queer people's formation of families of choice. Queers have always begat and reared children, but traditionally in different-sex marriage settings or in the wake of their break-up. The last generation has seen the formation of lesbian and gay families in which children have been conceived or adopted within the context of same-sex unions. The next generation will see the formation of lesbian, gay, and trans-sexual families whose children may sometimes be the consequence of cloning or gene-splicing technologies. In short, queer cloning has the potential to give queer people more ways to make families (thereby overcoming present restriction on their doing so) and allows for the further separation of reproduction and sex.

On the other hand, queer cloning or gene-splicing will in the short term be costly, especially for men, and risky for all queer people. Because family law has been conceived with heterosexual married couples in mind, it yields gaps and barriers to queer families generally, and families with clones in particular. In some jurisdictions, family law is already pervaded with antigay bias. Our fear is that even the possibility of queer cloning will trigger further antigay measures, and perhaps even anticloning measures as well. Nevertheless, we remain essentially optimistic that queer cloning will in the long run expand gay people's options for family formation, "normalize" queer people as more of them beome parents as well as partners, and perhaps even contribute in some modest way to the erosion of gender, sex, and sexual orientation as stigmatizing traits.

Endnotes

1. There has been considerable controversy about the use and meaning of the term *queer*. We use it as a general term for people who are not heterosexual in their sexual orientation.

2. See Kath Weston, *Families We Choose* (New York: Columbia Univ. Press, 1992), as well as Laura Benkov, *Reinventing the Family: The Emerging Story of Lesbian and Gay Parents* (New York: Crown, 1994); William N. Eskridge, Jr., *The Case for Same-Sex Marriage* (New York: Free Press, 1996); David Estlund and Martha Nussbaum, editors, *Sex, Preference, and Family* (New York: Oxford Univ. Press, 1997).

3. On alternative reproductive techniques, see, for example, John Robertson, *Children of Choice: Freedom and the New Reproductive Technologies* (Princeton: Princeton Univ. Press, 1994).

4. See Benkov, *Reinventing the Family;* William Eskridge, Jr., and Nan D. Hunter, *Sexuality, Gender, and the Law,* 827–867 (Westbury, NY: Foundation Press, 1997), surveying legal debates and obstacles to lesbian and gay parenting.

5. I. Wilmut *et al.,* "Viable Offspring Derived from Fetal and Adult Mammalian Cells," *Nature* 385 (Feb. 27, 1997), 810–813. See also Gina Kolata, "Rush Is on for Cloning of Animals," *New York Times,* June 3, 1997.

6. A transsexual is a person whose gender identity does not match his or her sex at birth *and* who has undergone or plans to undergo surgery to change his or her sex. To be contrasted are hermaphrodites, or intersexed persons, whose sex at birth is in some way or another ambiguous. Many transsexuals after surgery are infertile; some intersexed persons are infertile as well. In the discussion in text, we use the term *transsexual* to denote infertile transsexuals. See generally Bernice Hausman, *Changing Sex: Transsexualism, Technology and the Idea of Gender* (Philadelphia: Temple Univ. Press, 1995).

7. See *In re* Jacob, 660 N.E.2d 397 (N.Y. 1995); *In re* Adoption of B.L.V.B., 628 A.2d 1271 (Vt. 1993); Adoption of Tammy, 619 N.E.2d 320 (Mass. 1994); *In re* Adoption of Two Children by H.N.R., 666 A.2d 535 (N.J. Super. 1995). Contra, *In re* Angel Lace M., 516 N.W.2d 678, 683 and nn. 8–9, 11 (Wis. 1994) (unmarried mother's life partner could not adopt child without terminating mother's parental rights because statute literally required such termination unless birth parent is spouse of adoptive parent).

8. The "lesbian baby boom" has been going on since the 1980s, when many lesbians and lesbian couples decided to have babies and rear children in lesbian households. See Nancy Polikoff, "This Child Does Have Two Mothers: Redefining Parenthood To Meet the Needs of Children in Lesbian-Mother and Other Nontraditional Families," 78 *Geo. L. J.* 459 (1990).

9. According to M. V. Lee Badgett, "The Wage Effects of Sexual Orientation Discrimination," 48 *Indust. & Labor Rels. Rev.* 726 (1995), the annual average earnings of employees by sex and sexual orientation were as follows:

Heterosexual Men	$28,312
Homosexual/Bisexual Men	$26,321
Heterosexual Women	$18,341
Homosexual/Bisexual Women	$15,056

Badgett's data reflect the previously documented wage gap between women and men and suggest that a similar wage gap exists between gay/bisexual and straight employees for each sex. Admittedly, Badgett's findings are preliminary. Because the sexual orientations of the employees were based on self-reporting to the surveyors, it is hard to be sure the numbers capture the nuances of the workplace. For example, would the incomes of completely closeted but apparently queer employees reflect this disparity?

10. See note 9 above.

11. See Marla Hollandsworth, "Gay Men Creating Families Through Surro-Gay Arrangements: A Paradigm for Reproductive Freedom," 3 *Am. U. J. Gender & L.* 183 (1995), as well as Eskridge and Hunter, *Sexuality, Gender, and the Law,* 774–778, 849–853.

12. See, e.g., N.Y. Dom. Rel. Law § 122; *In re* Baby M, 537 A.2d 1227 (N.J. 1988).

13. D.C. Code § 16-401 (1993); Mich. Stat. Ann. § 25-248 (1994).

14. Fla. Stat. Ann. ch. 742.15, 742.16 (1994); Va. Code Ann. §§ 20-159 to -165 (1994). Arkansas is the only state that explicitly provides that an unmarried man can enter into an enforceable surrogacy contract with a woman. Ark. Code Ann. § 9-10-201 (1993).

15. *Johnson v. Calvert,* 851 P. 2d 776 (Cal. 1993).

16. This is the rule of the Uniform Parentage Act, § 5(a)-(b), which has been adopted verbatim in at least thirteen states. See Hollandsworth, "Surro-Gay Arrangements," 208 & n.110. Other states have a similar rule even though they do not otherwise follow the uniform law. See *id.* at 209 n.112.

17. *Jhordan C. v. Mary K.,* 224 Cal. Rptr. 530 (Cal. App. 1986).

18. See, e.g., Martha Field, *Surrogate Motherhood* (Cambridge, MA: Harvard Univ. Press, 1990); Elizabeth Anderson, "Is Women's Labor a Commodity?," *Phil. & Pub. Affs.,* Winter 1990, at 71; Margaret Radin, "Market Inalienability," 100 *Harv. L. Rev.* 1849 (1987); Debra Satz, "Markets in Women's Reproductive Labor," *Phil. & Pub. Affs.,* Spring 1992, at 107, all discussed in Eskridge and Hunter, *Sexuality, Gender, and the Law,* 779–780.

19. See, e.g., Lori Andrews, *Between Strangers: Surrogate Mothers, Expectant Fathers, and Brave New Babies* (New York: Harper & Row, 1989); Carmel Shalev, *Birthpower* (New Haven, CT: Yale Univ. Press, 1989), discussed and analyzed in Eskridge and Hunter, *Sexuality, Gender, and the Law,* 779–780.

20. J. Michael Bailey, Michael Dunne, and Nicholas Martin, "Sex Differences in the Distribution and Determinants of Sexual Orientation in a National Twin Sample," unpublished manuscript.

21. J. Michael Bailey and Richard Pillard, "A Genetic Study of Male Sexual Orientation," *Arch. Gen. Psychiatry* 48 (1991), 1089–1096; and J. Michael Bailey, Richard Pillard, Michael Neale, and Yvonne Agyei, "Heritable Factors Influence Sexual Orientation in Women," *Arch. Gen. Psychiatry* 50 (1993), 217–223.

22. Bailey, Dunne, and Martin, "Sex Differences in the Distribution and Determinants of Sexual Orientation in a National Twin Sample."

23. William Byne and Bruce Parsons, "Sexual Orientation: The Biological Theories Reappraised," *Arch. Gen. Psychiatry* 50 (1993), 228–239.

24. Byne and Parson, "Sexual Orientation: The Biological Theories Reappraised"; Edward Stein, *Sexual Desires: Science, Theory and Ethics* (New York: Oxford Univ. Press, in press).

25. On queers who parent, see Charlotte Patterson, "Lesbian Mothers, Gay Fathers and Their Children," in Anthony D'Augelli and Charlotte Patterson, editors, *Lesbian, Gay and Bisexual Identities over the Lifespan* (New York: Oxford, 1995), 262–290; also several essays in *Dev. Psychol.* 31 (January 1995) [special issue on sexual orientation and human development].

26. An historical explication of these mutually reinforcing anxieties is in William N. Eskridge, Jr., *Gaylaw: Challenging the Apartheid of the Closet,* chs. 1–3 (Cambridge, MA: Harvard Univ. Press, in press).

27. *Baehr v. Lewin,* 852 P.2d 44 (Haw. 1993). In December 1996, the trial judge, on remand from the Hawaii Supreme Court, ruled that the state had not carried its burden of justification and therefore could not constitutionally bar same-sex couples from civil marriage. As we go to press, the state has appealed that ruling, and an initiative that could ultimately override the state courts on this issue will be voted on later in 1998.

28. Defense of Marriage Act of 1996.

29. Mary Anne Warren, *Gendercide: The Implications of Sex Research* (Totowa, NJ: Rowman and Allenheld, 1985).

30. Edward Stein, "Choosing the Sexual Orientation of Children," *Bioethics* 12 (1998); William Byne and Edward Stein, "Ethical Implications of Medical and Biological Research on the Causes of Sexual Orientation," *Health Care Analysis* 5 (1997), 136–148; and Edward Stein, Udo Schüklenk, and Jacinta Kerin, "Scientific Research on Sexual Orientation," in Ruth Chadwick, editor, *Encyclopedia of Applied Ethics* (San Diego: Academic Press, 1997).

Sex and the
Mythological Clone

Wendy Doniger

A *recent* item from the Associated Press was briefly but pithily re-
counted by the *New York Times* (Wednesday, June 25, 1997) under
the headline, "Vatican Warns Against Cloning": "Human cloning
would not lead to identical souls because only God can create a
soul, a panel set up by Pope John Paul II has concluded. The Pontifical Academy of Life said the spiritual soul, 'the constitutive kernel' of every human created by God, cannot be produced through
cloning." This consideration may not rank very high on the list of
objections to cloning that are now raging in medical, legal, and
political circles, but it is very deep-seated indeed, and may be sub-
consciously fueling the other, more "rational" objections. Though
science has only recently learned how actually to produce clones,
mythology has imagined for millennia that doubles could be produced by the ancient counterpart of science—magic—and has
generally regarded it as a lousy idea. When medicine, confronting
the achievement of cloning, exclaims, all dewy-eyed, "O brave new
world," mythology mutters, " 'Tis new to thee." [Shakespeare, *The
Tempest* 5.1]

Just about a century ago, European mythology reacted to dramatic new medical advances by imagining various sorts of cloning,
all depicted as evil for many of the reasons that still animate the
Pontifical Academy of Life. Inspired by the earlier, classic Romantic text about the use of cadavers and galvanization to create a

human clone (Mary Shelley's *Frankenstein,* 1818), three other great
Gothic novels on this theme were published in the last decades of
the nineteenth century and remain, with *Frankenstein,* a part of our
own contemporary mythology of cloning. These novels speak of our
horror not of the ghosts of the past but of the ghosts of the future,
of the medical science that is yet to come; they speak of the production of morally (and hence physically) flawed doubles cloned
through the use of psychotropic drugs (Robert Louis Stevenson's
The Strange Case of Dr. Jekyll and Mr. Hyde, 1887), vivisection (H. G.
Wells's *The Island of Dr. Moreau,* 1896), and blood transfusion and
hypnosis (Bram Stoker's *Dracula,* 1897).

These novels were reacting to specific scientific advances that
galvanized (if I may use the term) British society in ways similar to
those in which the recent advances in cloning have affected our
own society. To us, now, these texts are classics with many spin-offs,
notably their many, many clones in film versions from Hammer and
Hollywood, some of which I consider here. But in their own day
they themselves were just reruns of much older ideas about splash-
ing around in and mucking about with what we would now call the
gene pool. Enhancing the power of the new terrors by analogizing
them to the ancient terrors expressed in myths, these novels kept
the old myths alive by giving them the transfusion, as it were, of
new scientific terrors. Myths, like vampires, are un-dead. In this
essay, I will attempt to track the myth of cloning back to its lair in
the mythological roots of European civilization, to excavate the
soil to which, like Dracula with his Transylvanian coffin, the myth
returns again and again to be revived.

The premodern mythology of cloning deals with two basic issues, often in separate narratives: One is eugenics, the active attempt to influence the human embryo, and the other is the problem
of the erasure of individuality, the challenge set by the confrontation of multiple copies of a human being. Many myths have imagined this second problem as happening to adult twins (who are, in
mythology, overwhelmingly male or male/female, seldom female)

and nontwins (both men and women). The narratives about men are entirely different from those about women, and there are, moreover, two different sorts of narratives told about all of them (twins and nontwins, male and female): stories about the people who actively produce clones, and stories about people who passively encounter clones that someone else has produced. Let us consider the various situations one by one.

Embryos: Paternal Insecurity and Resemblance

A surprisingly large number of male authors, in different cultures over many centuries, have believed that a woman who imagines or sees someone other than her sexual partner at the moment of conception may imprint that image upon her child, thus predetermining either its appearance or aspects of its character, or both. This belief in mind over matter—what you think about is what you get—is called by various names: maternal imagination (a woman's fantasy about something or someone who may not be physically present), impression (the mental reception, and transmission to the embryo, of a visual image that is physically present), or imprinting. Since the father, though less often, is said to participate in these processes too, it is probably best called parental imprinting.[1]

This concept is attested first in stories about animal husbandry, such as the Hebrew Bible story of Jacob, who put speckled rods in front of the ewes being covered by rams, so that they would bring forth speckled lambs (Genesis 30:25–43 and 31:1–12), and in Greek texts such as the *Knyegetika,* attributed to Oppian, a Syrian of the late second or early third century C.E., who veers from his main line, horses and hounds, just long enough to apply the principles of their breeding to humans, too.[2] The *Gynecology* of Soran, an authority on obstetrics who lived at the turn of the second century C.E. in Rome and Alexandria, also assumes a correlation be-

tween human procreation and animal husbandry; Soran tells of a
misshapen tyrant of the Cyprians who compelled his wife to look
at beautiful statues during intercourse and became the father of
well-shaped children; without drawing breath, he mentions horse-
breeders who place handsome stallions (real ones, not pictures) in
front of mares being covered by (presumably other) stallions.[3] In effect, the tyrant is cloning the statue. But how did the mere sight of
the real stallions or statues influence the quality of the offspring?

A lost and probably apocryphal text attributed to Empedocles,
a pre-Socratic poet (circa fifth century B.C.E.), is quoted by Aetius,
who asks: "How do offspring come to resemble others rather than
their parents? [Empedocles says that] foetuses are shaped by the
imagination [*phantasia*] of the woman around the time of conception. For often women have fallen in love with statues of men and
with images and have produced offspring which resemble them."[4]
Soran seems to have taken the folk wisdom recorded by Empedocles, that women *do* fall in love with statues, and connected it with
the folk wisdom of animal husbandry recorded by the Hebrew
Bible (and elsewhere), that females can be *made* to desire obstetri-
cally, as it were, the images that the husband desires eugenically; in
the process, he has moved from the herd animals favored in the
Bible (sheep and goats) to the favorite animals of the Greeks,
horses. The result is an active attempt by the husband to treat his
wife like a mare (or a ewe): He shows her pictures of what he
wants her to give birth to.

This sort of mythological embryology involves a kind of prescientific cloning: It investigates ways of producing copies of desired
stock. But, we must ask, desired by whom? One factor that seems
to pervade all the ancient texts on this subject is the male desire to
control female desire. For when the positive approach to eugenics
on the animal husbandry model, in early Greek texts, was reintro-
duced into later Jewish and Christian texts, it was soon overwhelmed by male paranoia. In comparison with the earlier texts on
animal husbandry, these later texts about human embryology deal

largely with the fear of *negative* cloning: the influence of some man other than the father, to make the child *fail* to resemble the father. As long as the father controlled the process, he regarded it as a good idea; but when the mother had ideas of her own, the male authors of our texts generally regarded it as unacceptable.[5] The fear that his son will not look like him drives him to make sure what his woman is looking at, in the hope that she will not imagine some other man. It comes down to the matter of who initiates the fantasy: Her fantasy is only acceptable if it is, in fact, *his* fantasy, his idea of what she should be seeing while he makes love to her; and it is certainly easier to regulate external vision than internal vision.

Premodern theories of parental imprinting, motivated primarily by concerns about paternity and inheritance, usually assume that the male child (the only sort of child the authors of these texts cared about) will normally resemble its father (and, sometimes, to some extent, its mother). If he happens to resemble his father's best friend, some texts drew what seems to us the most obvious conclusion, namely, that the mother was sleeping with the father's best friend. Other texts, however, imagined that she might merely have *thought* about him, or looked at a *picture* of him, and that this was sufficient to imprint that false image on the baby's face. But the repressed knowledge of adultery is always there, and bursts out in various paranoid forms. Thus, Jewish midrashic texts, dating from the third or fourth century C.E., regarded maternal imprinting, whether by pictures or just by the woman's fantasy, as a form of mental adultery, better than physical adultery,[6] but still bad enough to brand the child a kind of bastard.[7] To men who fantasized that mental acts influenced the quality of their offspring, the very survival of the species depended upon the sexual fantasies of their women. Thomas Laqueur states the case very well: "Since normal conception is, in a sense, the male having an idea in the woman's body, then abnormal conception, the mola, is a conceit for her having an ill-gotten and inadequate idea of her own."[8] The eye was therefore an instrument of eugenics, passively and externally receiving the images produced by the father, while the mind (hidden inside the woman) was an instrument of adultery.

As if this were not bad enough, the theory was turned on its head to show that resemblance, too, could dissemble. In the seventeenth century in Europe, it was argued "that a woman who thinks strongly about her husband in the midst of illicit pleasures can produce, through the force of her imagination, a child that *perfectly resembles him who is not the father . . . Resemblance is not proof of filiation . . .*"[9] Now, it is easy enough to imagine that a woman might dream of her lover while in the embrace of her husband, but why, one might ask, would she think of her husband while in the embrace of her lover? One answer is that if she, too, subscribes to the theory of maternal imprinting, she will think about her husband when she is with her lover on purpose, in order to conceal her adultery. Thus, the mother controls her own imagination and produces "not a monster but its exact opposite: a child who actually resembles the legitimate spouse who did not father it."[10] This aspect of resemblance furnished yet another black mark in the copybook of cloning: a son who resembled his father *too closely* posed a threat, sometimes projected from the supposed cuckolder onto the son himself in stories in which a woman mistook her too-resembling son for her husband and slept with him, the old Freudian nightmare.[11]

The backlash from this switchback was very serious. For where the theory of maternal imprinting doubtless saved the necks of a number of adulteresses whose children did not resemble their fathers, the corollary (that an adulteress could imprint her lover's child with her husband's features) cast suspicion upon *all* women, indeed particularly upon faithful women, whose children *did* resemble their legitimate fathers.[12] Women as a whole were cast as body snatchers, who could at will replace a seemingly normal child with a monster conceived in the pods of adulterous beds. This was truly a no-win situation; women were damned if they did, and damned if they didn't, produce children who resembled their husbands. And the cognitive dissonance that resulted from this uncertainty drove men both to accelerate their attempts to control women's sexuality and to project their images, obsessively, upon their sons. Thus M. Boursicot, the French diplomat in the real-life affair that inspired David Henry Hwang's *M. Butterfly,*[13] explained why he thought his (male) Chinese lover was a woman and that she had borne him a son; speaking of that son, he said, "He looked like me."[14] Since cloning is about reproducing humans, which is to say it is about sex (or possible substitutes for sex), it is no wonder that sexual fears play such a large part in the mythology of cloning.

Male Twins: Identity and Resemblance

The scenario of parental imprinting assumes that the other man whom your son might resemble is either your rival (in negative cloning) or someone who will improve what we would call your genetic stock (in positive cloning). But this situation, nervous enough, is exacerbated by a fear not of nonresemblance but of resemblance, the fear that your rival—in flesh or in a picture—will *look like you,* and that your wife will mistake him for you and therefore betray you with him. Thus, fears of both nonresemblance and resemblance are keyed to the father's desire to imprint his child himself, usually with his own image; the interference either of the mother's imagination of another man (the problem of nonresemblance) or of the physical or mental presence of another man who resembles the husband is a threat to paternity and inheritance.

The most obvious natural clones are identical twins, who play a major role in many of the creation myths of the world, often negative; many premodern cultures believe that one, or both, of twins should be killed at birth. Of the many aspects of this mythology, the one that concerns my argument here is the problem of sexual identity. In Plautus' *Menaechmi,* composed in the early second century B.C.E., Menaechmus unwittingly masquerades as his brother and enjoys both his brother's dinner and his courtesan. In another play by Plautus, *Amphitryon,* the natural cloning of twins is replaced by the magic cloning of two gods (Jupiter and Mercury), who magically assume the forms of Amphitryon and his servant, Sosia, so that Jupiter can seduce Amphitryon's faithful wife. The story of Amphitryon was retold a number of times, notably by Molière (1668), Heinrich von Kleist (1807), Jean Giraudoux (1929), and S. N. Behrman (1938). Shakespeare combined and adapted both plays to produce his great slapstick tale of twin brothers *and* their twin servants, no longer doublets but quartets: *The Comedy of Errors.* The innocence and comedy inheres in the fact that, like the brothers in *Menaechmi* rather than the gods in *Amphitryon,* the double, the clone, *does not intend* to seduce his twin brother's wife; on the contrary, the joke is that he does *not* desire her as she expects he will. Here is a brief summary of the plot:

> Twin brothers, both named Antipholus, were separated in youth as the result of a shipwreck; one lived in Ephesus (E. Antipholus, henceforth E.), the other in Syracuse (S. Antipholus, henceforth S.), and each had a servant named Dromio (E. Dromio and S. Dromio). E. was married to Adriana but S. was not attached. One day, S. came by chance to Ephesus, with S. Dromio, and was mistaken for E.

The sexual "errors" turn upon the question of sexual jealousy and rejection: Adriana wonders why S. (her husband, she thinks) does not love her, and, indeed, her worries are not unfounded: For E. is suspiciously friendly with a courtesan, as he himself admits, and Adriana worries that she is growing old and losing her husband's love. Thus, where the wife (rightly or wrongly) suspects what we have come to call "the other woman," she begins to believe that her husband actually is "the other man"—which he, or rather his twin brother, really is. Similarly, the unattached S. Dromio is puzzled to discover that E. Dromio's girlfriend, who of course seems to him to be a total stranger, regards him as her lover.

The Comedy of Errors is about dissymetry in sexual love: One partner expects intimacy, while the other regards him or her as a
stranger. When Adriana mistakes S. for her (straying) husband, she
says, "I am not Adriana, nor thy wife," and explains that since she
is part of him, and he is false, she is false. "I know you not," he
replies, and means it literally (in both senses of "know," intellectual
and carnal knowledge), while she thinks he is merely speaking
metaphorically, as she was. Adriana says that for weeks E. has been
"much, much different from the man he was." (5.1) And now he is
really really different. The unusual accident of double twins triggers
the not so unusual accident of marital infidelity and estrangement:
The feeling that one simply *does not recognize* one's official partner
is expressed in the dream (or nightmare) that that partner actually
is a total stranger.

The Comedy of Errors was adapted, centuries later, a number of
times, into various Gilbert and Sullivan operettas and the musical
comedies *The Boys from Syracuse, A Funny Thing Happened to Me on the
Way to the Forum,* and *Don't Start the Revolution Without Me*). The
theme is a perennial favorite. It even inspired a female variant, in
Big Business (1988, directed by Jim Abrahams), in which both Bette
Midler and Lily Tomlin play identical twins:

> The Midler twins are born of a couple from New York, and the
> Tomlin twins of a couple from a small Tennessee town, simulta-
> neously in the same hospital. A nurse switches one of each, in the
> Gilbert and Sullivan manner, so that one Midler and one Tom-
> lin is left with each set of parents. The Midler twin in the coun-
> try and the Tomlin twin in the city have a sense that they are
> misfits who belong somewhere else; the Midler twin in the city
> and the Tomlin twin in the country, on the other hand, are suc-
> cessful in their work and with men, though they break away
> from them (the city Midler divorces her husband and the coun-
> try Tomlin abandons her boyfriend). Eventually the twins meet
> in New York, and the men who have been rejected by the suc-

cessful twins are encouraged by the misfit twins, but since no one knows that there are doubles, the stop-go, hot-and-cold effect drives the men crazy. Finally they switch back to where they should have been, and the misfit twins end up happily with the men who had been rejected by their successful twins, while the successful city Midler finds a new man and the successful country Tomlin accepts the man rejected by the misfit city Tomlin.

Now country mouse and city mouse have been added to the mix; when the women meet their twins, the city Midler says scornfully, "It's me with a bad haircut," and the city Tomlin moans, "I hope I'm me." The country Midler shrieks, "They're pod people! I saw the movie!" to which the country Tomlin replies, "You watch too many movies." The Shakespearean problems remain: The men don't know why they have been accepted or rejected, and they offer a modern version of the old answer to the question of inconsistent (female) behavior: "Do you suppose it's hormonal?"

The many variants of the tale of twin brothers, from ancient folklore to the contemporary cinema, can be divided into two main groups: those in which (as in *Menaechmi* and *The Comedy of Errors*) one twin is unaware of the other's existence, unwittingly masquerades as the other, or rejects the other's beloved; and those in which (as in *Amphitryon*) one twin purposely masquerades as the other in order to bed the other's beloved. In the first group we might include the many variants of the folk tale of "The Two Brothers," best known to us in the variant recorded by the brothers Grimm,[15] in which the double lays a sword in the bed between himself and his twin brother's wife. (The sword motif recurs in stories not about twins but about magical doubles—like Jupiter/Amphitryon—in the Welsh story of Pwyll in the *Mabinogion*[16] and in the story of Siegfried, recorded in Norse mythology,[17] and in Richard Wagner's opera, *Götterdämmerung;* a variant also occurs in African mythology).[18] The theme of the unwitting masquerade remained popular in Hollywood, where Danny Kaye practically cornered the market on male twins who find themselves innocently mistaken for their doubles: *Wonder Man* (1945), *On the Riviera* (1951), *On the Double* (1961). And in the same year—1944—that Betty Hutton sang duets with herself in *Here Come the Waves,* Gene Kelly danced with his alter ego in *Cover Girl.*

In the second group, where the seduction is intended, we might include the folk variants listed by Stith Thompson under the motif of a twin who deceives the wife of his brother (motifs K 1915–1917, K 1311);[19] E. T. A. Hoffmann's "The Doubles" (*Die Doppelgänger,* 1821); *Lives of the Twins* by Rosamond Smith (Joyce Carol Oates, 1987); and many films, from *The Corsican Brothers* (1941), in which Siamese twins, separated shortly after birth, later become bitter rivals when one falls in love with the woman who loves the other, to the Canadian film *Dead Ringers* (1988), based upon the 1975 suicide of two real twin gynecologists, Cyril and Stewart Marcus, who often impersonated one another with women.

Male Clones: The Double Bind

Men, or male gods, who are not natural twins may use magic to make themselves into clones of other men. This happens in two sorts of narratives, really the same story told from two different points of view. Sometimes a man or a god (like Jupiter) will produce a double of himself in order to get to the woman he wants (the woman who then may react by splitting off *her* clone). But sometimes a man suddenly encounters a double he didn't know he had (as Amphitryon encounters Jupiter in the form of Amphitryon). In the European retellings, the emphasis shifts from the motives of the cloner to the reaction of the clonee, the man who suddenly confronts a double he did not know he had. The Amphitryon reaction lies at the heart of our own adverse reaction to cloning: the gut feeling that there should not be two identical forms of anything, especially of me.

Let us take another look at Amphitryon. The name of his servant, cloned by Mercury, is Sosia, which comes to mean "a double" in both French and Italian; in French, *sosie* means "a person who bears a perfect resemblance to someone else."[20] Robert's French dictionary, citing occurrences in J. de Rotrou's *Les Sosies* (1638) and Molière's *Amphitryon* (1668),[21] summarizes this character in Plautus: "Sosia finds himself face to face with another Sosia. . . . He comes to doubt his own identity." Sosia becomes hopelessly confused and finally asks the god Mercury (in the form of Sosia), "Then who am I, if I'm not Sosia, will you kindly tell me that?" To which Mercury replies, "Well, when *I* don't want to be Sosia, *you* can be Sosia yourself."[22]

Different retellings of the Amphitryon story raised different questions about the sense of self threatened by the confrontation with the clone, not only in the humans—Sosia, Amphitryon, and his wife Alcmena—but even in the god, Jupiter. For Jupiter experiences the usual paradox of the double bind that plagues all intentional doubles who engage in sexual masquerades: You disguise yourself in order to extend your powers, to get into bed with someone else's woman, but then it isn't you she loves, but the other, of whom you are insanely jealous. Jupiter keeps asking Alcmena, "Wasn't the sex better last night than on other nights?" but she resolutely insists that she couldn't tell the difference—which was, but also was not, what Jupiter intended.

The myth of the sexual double is often set in the context of some other (often related) concern that a particular culture regards as even more basic than sex, such as fraternal succession or political rivalry. Many clones are about death; Otto Rank summarizes well the theme of the double in literature of the nineteenth-century Romantics, such as Hoffmann:

We always find a likeness which resembles the main character down to the smallest particulars, such as name, voice, and clothing—a likeness which, as though "stolen from the mirror" (Hoffmann), primarily appears to the main character as a reflection. Always, too, this double works at cross-purposes with its prototype; and, as a rule, the catastrophe occurs in the relationship with a woman, predominantly ending in suicide by way of the death intended for the irksome persecutor. In a number of instances this situation is combined with a thoroughgoing persecutory delusion or is even replaced by it, thus assuming the picture of a total paranoiac system of delusions.[23]

In stories like Edgar Allan Poe's "William Wilson" (1840), Théophile Gautier's *Avatar* (1856), and Maurice Renard's *Le docteur Lerne, sous-dieu* (1919), the protagonist, intent on seducing another man's wife, encounters his double and, in killing him, kills himself.

Other nineteenth-century clones were bureaucratic (the mindless, faceless doubles that run the world), like Golyadkin, the pitiful protagonist of Dostoyevski's *The Double* (1846); but Golyadkin loses to his bolder and more successful double not only his job but also the woman whom he loves from afar.[24] So, too, a number of nonmagical clones were produced, with the mere application of a bit of makeup, for the purposes of spying, or of protecting a powerful political figure. Alexander Dumas's novel *The Man in the Iron Mask* (1847), in which King Louis XIV has a twin who replaces him, was cloned several times, in films by Douglas Fairbanks (1929) and Warren William (1939). The novel was cloned in a book by Sir Anthony Hope, *The Prisoner of Zenda* (1894), in which an Englishman takes the place of a "Ruritanian" monarch whom he resembles exactly (though they are only distantly related). This, then, was cloned in several films, a serious swashbuckler with Ronald Colman (1937), a satire by Peter Sellers (1979), and several more distant relatives: *Moon Over Paradour* (with Richard Dreyfuss, 1988) and *Dave* (with Kevin Kline as the clone of the American president, 1993). In each of these, despite the political premise, the identity crisis occurs in the bedroom, when the princess or First Lady tempts the clone to give up his true identity forever.

Then there are the science fiction clones in films, beginning
with a great political film of the silent era, Fritz Lang's futuristic
Metropolis (1926), in which a robot clone of the heroine fools everyone except the hero who loves her. In *Invasion of the Body Snatchers*
(1956), seed pods from outer space land in a small town and begin
to replace its inhabitants one by one with mindless clones. Their
goal is political, as is the agenda of the filmmaker: The assumption
that the clones have no emotions, no feelings at all, is a clue to the
movie's function as a thinly veiled anticommunist tract, in which
the possessed hand over their minds to Mars, aka Moscow; the anticommunist party line traditionally depicted communists as emo-
tionless (recall *Ninotchka* [1939]). But the test of identity is still
sexual: When the hero's girlfriend has been replaced by a clone, he
does not realize this until he kisses her, an act that he later recalls
with the immortal line, "I've been afraid a lot of times in my life,
but I didn't know the real meaning of fear until I'd kissed Becky."
And the cloning becomes explicitly sexual in *The Stepford Wives*
(1974, written by William Goldman from the novel by Ira Levin),
in which suburban husbands replace their wives with computerized
robots who love to clean house and praise their husbands' sexual
performances, and who lose their desire to do anything that makes
them individuals. That is, the clones are created precisely for their
mindless sexuality.

The Manchurian Candidate (1962, from the novel by Richard Con-
don) used vaguely "Commie" brainwashing techniques to implant
false memories in Lawrence Harvey; *Blade Runner*, directed by Ridley Scott in 1982, from Philip K. Dick's short story "Do Androids
Dream of Electric Sheep?" used more complex and explicitly clone-
like techniques (quickly satirized, in 1983, in *The Man with Two
Brains*, in which Steve Martin plants the disembodied brain he loves
in the gorgeous body of his despised wife, Kathleen Turner). False
identity, more precisely false memory, remained the basic issue in
later films about scientific clones, like *Total Recall* (1990) and *Duplicates* (1992), in which evil surgeons implant false memories in

the brains of innocent people. In another branch of this genre, it is the face, rather than the brain, that is cloned, producing a more superficial identity crisis. In *Shattered* (1991), the amnesiac victim is given both memory and (through plastic surgery) the face of another man; since he is still loved by the woman who had loved *him,* presumably his memory and his face, the film asks, if you don't have your face or your memories, who are you? *Face/Off* (1997) simplifies the problem: The memory remains, and only the faces are transferred from one man to another. Despite the political basis of most of these films (*Total Recall* is about a political tyranny and resistance on Mars; *Duplicates* about a government plan to replace the memories of socially undesirable people; *Face/Off* about an FBI man and the terrorist he is hunting), the crunch still usually comes in the bedroom, where the victim regains a kind of physiological memory that has escaped the quasi-cloning process.[25] In *Face/Off,* the terrorist, with the face of the FBI man, successfully fools the FBI man's wife in bed; when he tells the people at the FBI how to find the terrorist, and they get suspicious and ask, "How do you know so much about this guy?" he replies, "I'm sleeping with his wife" (which is true only of the invisible self beneath the face—the terrorist, not of the apparent self in the face—the FBI man).

Many of these themes combine in a hilarious satire on the theme of cloning, the 1996 film *Multiplicity,* in which the twin who tries not to sleep with his brother's wife fails hilariously, several times:

Doug (Michael Keaton) wants more time to do the increasing number of tasks demanded of him in his work as an engineer and more "quality time" with his kids and his wife Laura (Andie MacDowell), who wants to go back to her job in real estate. He lets a geneticist create two clones of himself (Doug 2 and Doug 3), and Doug 2 on his own produces a clone of himself, Doug 4, who is, like the blurry photocopy of a photocopy, an idiot. Doug warns the clones away from his wife (Rule Number One), but one night when Doug is away, though Doug 3 loyally puts a pillow between them in the bed, Laura aggressively and irresistably seduces first him and then the other two clones.

Laura notices the discrepancies, memory lapses, and so forth; "I feel like I don't know you anymore," she says, like the wife of the twin in *The Comedy of Errors*. But she still thinks there is just one of him. And when she threatens to leave Doug in fury at something one of the clones did, he protests, "It wasn't me," invoking the split-level alibi that we will encounter below in classical myths about raped women.

At one point, one of the clones suggests that they "clone Laura," but in fact it is because Laura is already about to do the opposite of cloning, because she is about to split (into mother and real-estate agent), that her husband in desperation doubles himself (and then, to outdo her, trebles and quadruples himself into the mythological lineup). At the end, he defends himself by accusing her: "It's happened to you, too. You want to be a mom but you also want to work." Moreover, in justifying the infringement of Rule Number One, the clones say that she was "unstoppable," "a very powerful woman"; they suggest that she is insatiable (an idea confirmed by our own witnessing of the three seduction scenes). Believing that they must resist her, the magical doubles invoke the mythological technique of the sword (here a pillow) laid between them in the bed, but to no avail.

Doug has to get rid of them because he finds himself in the usual double bind of the man who both does and does not want his double to sleep with his wife: He is jealous of the clones, and feels that he has not doubled himself, as he hoped, but rather halved himself. When he finds out that Rule Number One has been broken, and reacts with shocked fury, the clones reassure him: "She thought I was you"; "She thought *I* was you too." And then: "We're not perfect." This last statement ostensibly means just, "We were susceptible to her," but in this context it also means, "We are not perfect copies; you alone are you." Now Doug asks them, as Jupiter asked Amphitryon's wife Alcmena, "Do you think she liked you more than she likes me?" "Of course not," Doug 2 says, "I *am* you; you're me."

But since this clearly does not address the problem, Doug 3 (the sensitive one) assures him, "I know she loves you, Doug. I wasn't really *there* for her. Even when I was there, I wasn't *there*." The psychobabble cliché is here cunningly applied to a liminal situation in which "there" means simultaneously "existentially real," "sexually intimate," and "orgasmically on the G-spot." But the best line of all, the one that sums up the double bind in a phrase worth hundreds of pages of Freud and Otto Rank, is what Doug calls Rule Number One: "No clone nookie; original nookie only."

Female Twins: Good and Evil

Female twins, too, can play these tricks, often, like male twins, in order to commit adultery—betraying their twin sisters, rather than their husbands (their victims in the scenario of imprinting); but they are worse than male twins, because they lack the balancing scenario of the loyal twin (with the sword in the bed) that marks the mythology of male twins. Prepubescent twin girls can be quite charming and even useful (like the two Hayley Mills characters in *The Parent Trap* [1961], and even the Doublemint Girls who double your [unspecified] fun), but when they grow up and become sexually active (and threatening), they usually become murderous. (The Hayley Mills twins remained essentially prepubescent; twenty-five years later, in *The Parent Trap: II* [1986], the grown-up Mills replaced her sister on an innocent date with a man she wanted to win for her shy sister.) The folk theme of the "Substitute Bride" or "The Black and White Bride" (Stith Thompson's Tale Type 403), which often involves twin sisters, fuels an entire genre of Hollywood films of "the good and evil twin," that Jeanine Basinger calls the "My God! There's two of her!" theme (from the poster for *Dark Mirror*).[26] ("There's two of them!" says the villain who first sees the two Jean-Claude Van Dammes together in *Double Impact* [1991]). Special effects allowed the grande dame to play two parts simultaneously on the screen, resulting in all of those melodramatic tours de force such as *The Divorce of Lady X* (1938, Merle Oberon) and *Two-Faced Woman* (Greta Garbo's last film, made in 1941). So widespread is the film scenario that when a real woman impersonated her real (and murdered) sister in 1997, the police chief remarked, "I don't know if bizarre quite describes it; this is the stuff of a Hollywood movie."[27]

Filmes noires often depicted the evil twin as a successful murderer, such as *The File on Thelma Jordon* (1949, Barbara Stanwyck), which ends with the dying woman saying, "In a way, I'm glad it's over, the struggle, the good and evil. Willis [her lawyer] said I was two people; he was right. You don't suppose they could let just half of me die?" Bette Davis and Olivia de Havilland competed for the crown of queen of the good-and-evil twins, beginning in a film, *In This Our Life* (1942), in which Davis and de Havilland played together, not as twins but as sisters: Davis, the bad sister, stole the man who belonged to de Havilland, the good sister.

In subsequent films, Davis and de Havilland played separately, and each became both the good twin with the guy and the evil twin who took him away. In 1946, Olivia de Havilland made *The Dark Mirror,* which ends with the psychiatrist saying, only half jokingly, to the remaining, good, identical twin, "Why are you so much more beautiful than your sister?" In that same year, Bette Davis made *A Stolen Life* (the remake of a 1939 film with Elisabeth Bergner), in which, as usual, the evil Davis steals the good Davis's fiancé; the twist comes when the masquerader learns that the envied original was not in fact loved (a cinematic surprise already known from the film of Daphne Du Maurier's *Rebecca,* in 1940), so that she has gained nothing by the masquerade. As one critic remarked of this film, "What I'm waiting for is a film about beautiful identical quintuplets who all love the same man."[28]

Davis stamped the genre as her own by starring in the clone of this film almost twenty years later: *Dead Ringer* (1964), this time involving murder as well as sexual betrayal: The evil twin is killed by the good twin, who gets away with it but then is hanged for the murder that her sister had committed (the murder of the man that the good twin loved). This is another variant of the theme of discovering that the one whom you clone in fact has a worse life, not a better life, than you do. *Killer in the Mirror,* made in 1986 (with Ann Jillian), appears, at first, to be yet another clone of the old Bette Davis film, but it adds a new twist: The evil sister is not in fact dead, but pretends to be dead—indeed, engineers the whole switch of identities—in order to take over the life of the good sister in order to avoid hanging for the murder that (as in *Dead Ringer*) she herself committed.

In Hollywood films about twin sisters made in the 1990s, too, murder often played the leading role. Sometimes the women are not twins but unrelated women, one of whom turns herself into a clone of the other and commits a murder for which the other is suspected. This is the case with a number of films ranging from Alfred Hitchcock's classic *Vertigo* (1958), in which Kim Novak masquerades as a woman she has in fact helped to kill, to *Basic Instinct* (1991), in which Jeanne Tripplehorn masquerades as Sharon Stone when she commits a murder. The same thing happens in *Single White Female* (1992), made from a novel by John Lutz whose excellent title, *Single White Female Seeks Same,* expresses the longing for sameness that characterizes both urban paranoia and sexual paranoia.

Some contemporary films about twin sisters are about politics, rather than murder. In a Hungarian film, *My Twentieth Century* (directed by Enyeko Ildiko, in 1990), there are twin sisters—one, in the West, sexy and given to luxurious tastes like diamond necklaces and champagne; the other, in the East, poor and dowdy and revolutionary, given to throwing bombs. A clone of *My Twentieth Century* is the Polish/French film *The Double Life of Véronique,* issued just one year later (1991), about two women, Véronique and Weronika

(Irene Jacob), born at precisely the same time, one in France and one in Poland. In these films, the twins are separated not by a magic mirror but by an Iron Curtain, yet here, too, they invade one another's sex lives. Finally, and predictably, the story of the good-and-evil twins generated at least one pornographic double, *Mirror Images* (1991), in which Delia Sheppard plays a promiscuous and sexually repressed set of twin sisters (a distinction that also characterized some of the good-and-evil sisters), who replace one another in various beds belonging to men who do not really seem to care who they are in bed with; the surviving sister sums it up with a classic sexist remark in the last line of the movie: "Does it really make any difference?" This is the bottom line in female cloning: All women are alike, anyway.[29] Indeed, the assumed genetic transparency of women, the assumption that a woman is just a man's way of making another man, just as a chicken is just an egg's way of making another chicken, is what made maternal imprinting a problem for men in the first place.

Female Clones: Sexual Abuse and Flight

Women in the mythologies of female twins thus share the evil propensities of male twins, and then some. But, just as male clones took two basic forms, the *Menaechmi* (inadvertent masquerade by natural twins) and the *Amphitryon* (intentional masquerade by artificial clones), females too have another sort of clone in addition to natural twins. But the female clone of this type is the very opposite of the *Amphitryon* model: Whereas males clone to win something, inevitably to the loss (or, occasionally, destruction) of their rivals and often of themselves as well, the female clone in this scenario is produced in order to avoid, rather than to steal, a lover— or even a husband. And despite the male bias of the texts in which these narratives appear, the stories themselves often express sympathy for the women forced to act in this way.

In the *Rig Veda* (circa 1000 B.C., in northern India) a woman
who cannot stand her husband, or a stranger who is attempting to
rape her, goes away and leaves an identical clone in her place, her
shadow or reflection; in ancient Greek mythology, too, a phantom
stands in for Helen when Paris comes to carry her off to Troy.[30] (In
later, more realistic texts, the woman may just send her sister, or a
servant—as Isolde sends Brangane, in *Tristan and Isolde).*[31] Stories
such as these may express a kind of dissociation in reaction to a
rape: "This happened to some other woman, not to me." The clone
implies that the "real self" did not experience the event.[32] These are
stories of denial, and of the asymmetry even in many less violent
sexual acts, where one partner is "there" and the other is not. Thus,
the meaning of the splitting as perceived by the person who clones
is not merely "I am one, but I am also two," but "This is happening
to me and this is not happening to me," and "I am here and I am not
here." The mind/body separation experienced by victims of rape
is often invoked in contemporary discussions of sexual violence.[33]
In myths of rape, the clone serves to exonerate the woman herself
from any possible defilement at the hands of the demonic rapist or
unwanted husband.

Sexual aversion may lead to maternal imprinting, too: The animal husbandry model of Jacob's rods model that we considered
previously works simply enough in the European folk belief that a
woman frightened by the sight of a deformed man or a knife will
bring forth a deformed child or a child with a birthmark in the
shape of a knife.[34] But what if that "thing" or "object" is her husband?
Jewish and Christian texts do not seem to have devoted much attention to this possibility, but an important ancient Indian myth tells
of a woman, forced to accept a man she finds sexually loathesome,
who closes her eyes and therefore brings forth a blind child. Her
sister, subjected to the same man, takes the realistic-clone measure
and sends her serving girl in her place, thereby virtually equating
the uses of maternal imprinting and female cloning.[35] Just as the sisters in this story, forced to be impregnated by a man who disgusts

them, take measures to avoid him, measures that imprint the resulting children, so too European patriarchy and the *droit du seigneur* force women to be impregnated by men they don't want—and therefore, according to the theory, to imagine men they do want and to imprint the resulting children.

The women in these texts have limited options, but within them they have agency, an instinct for self-preservation, and a desire for sexual choice that the narratives seem to validate. To a female, let alone a feminist, reader, these stories depict women in a more positive light than the stories of female twins, or the tales of maternal imprinting and male cloning (which usually pay no attention at all to the women who are the objects of the male clone's desire). But from the standpoint of a male reader (or, more to the point, author), these are evil women: They abandon their husbands (and, as the texts make clear, their children). Thus, through either seduction or rejection, the female clone poses a sexual problem for men.

Conclusion

Each of these prescientific clones (embryos, twins, men, women) turns out to be at the very least nasty, at the very most totally destructive; mythology comes down strongly against cloning. The case against cloning is made over and over; even the comedies have a tragic aspect, and the tragedies are grim indeed. Many of these stories of clones are about things other than sex—they are about trying in vain to cheat death, and discovering one's identity, and politics, and (as the Pope says) about the uniqueness of souls. But time and again one clone somehow or other stumbles in the other's bed, and this feel of advertent or inadvertent sexual betrayal is, I think, an inescapable part of the terror of cloning.

The Pope's argument is deeply embedded in these myths; though more secular critics might use another word than *soul* to designate "the constitutive kernel" of a person, a word like *personality,* or *self,*

or *ego,* or *memory,* the myths resist the idea of two identical souls (or
their equivalents) in two identical bodies. (Some contemporary
Buddhists apparently wonder whose karma the clone would
have.)[36] Christian missionaries in Africa argued that identical twins
had two different souls, one each, and the missionaries' intervention did much to curtail the African practice of murdering one or
both twins at birth.[37] Yet the very myths that assume that identical
twins may have very different personalities (or souls) resist the idea
that a *clone*—that is, a magically created double—could have a separate soul, that in creating a body de novo, the magician could create a soul de novo too. We, too, instinctively feel that the soul is
inextricably linked to the body, and so if we see someone else with
our body, we presume that he or she must have taken something of
our soul. The question that we ask the mythological clone is, "If you
are me, who am I?"

Endnotes

1. See Wendy Doniger and Gregory Spinner, "Misconceptions: Female Imaginations and Male Fantasies in Parental Imprinting," in *Daedalus,* in press.
2. Oppian, *Kynegetica,* 1.327–8. Text and translation in *Oppian Colluthus Tryphiodorus*
   (translated by A. W. Mair; Loeb Classical Library, vol. 80; New York: G. P. Put-
   nam's Sons, 1928), pp. 34–5.
3. Soran, *Gynecology* 1, par. 39; text in *Soranos d'Éphèse: Maladies des Femmes* (text and
   translation by Paul Burguière and Danielle Gouryevitch; Paris: Les Belles Lettres,
   1988), p. 36; see also translation by Oswei Temkin, *Soranus' Gynecology* (Balti-
   more: Johns Hopkins, 1956), pp. 37–8.
4. Empedocles, cited by Aetius 5.12.2, *Doxographi Graeci* (edited by Herman Diels;
   Berlin: Walter de Gruyter, 1965), p. 432. See also *The Poem of Empedocles* (text and
   translation by Brad Inwood; Toronto: University of Toronto Press, 1992), p.
   185.
5. An important exception is offered by Paracelsus, writing in the sixteenth century
   in Germany, who granted that a woman's imaginations, too, could affect the embryo in positive—intellectually as well as physically positive—ways: If, at the moment of conception, she imagined a learned wise man, such as Plato or Aristotle,

or a warrior, Julius or Barbarossa, or a great artist, like the painter Dürer, she would bear a child like him. Paracelsus, *De Morbis Invisibilis,* in Hans Ranser, editor, *Schriften, Theophrasts von Hohenheim gennant Paracelsus* (Leipzig: Insel Verlag, 1921), Section 202, "Wirkung der Imagination," pp. 314–15.

6. Bemidbar Rabba 9:34; edited by M. A. Mirkin, *Midrash Rabba* (Tel Aviv: Yavneh, 1977), pp. 213 ff.

7. Kallah 50b, Kallah Rabbati 52a; Isaiah Tishby, *The Wisdom of the Zohar: An Anthology of Texts* (3 vols; translated by David Goldstein; Oxford University Press, The Littman Library, 1991 [1949]), vol. 2, pp. 646–9.

8. Thomas Laqueur, *Making Sex: Body and Gender from the Greeks to Freud* (Cambridge, MA: Harvard University Press, 1990), p. 59.

9. Marie-Hélène Huet, *Monstrous Imagination* (Cambridge, MA: Harvard University Press, 1993), pp. 79–80.

10. *Ibid.,* p. 80.

11. See the ancient story of Indra and Kutsa, *Jaiminiya Brahmana* (edited by Raghu Vira and Lokesha Chandra; Series 31; Nagpur: Sarasvati-vihara, 1954) 3.199–200; Wendy Doniger O'Flaherty, *Tales of Sex and Violence: Folklore, Sacrifice, and Danger in the Jaiminiya Brahmana* (Chicago: University of Chicago Press, 1985), pp. 75–6.

12. *Ibid.,* p. 81.

13. David Henry Hwang, *M. Butterfly* (New York: Plume [Penguin], 1989).

14. Eric Gerber, "Not-so-hot a Lover," *Houston Post,* Wednesday, May 21, 1986.

15. "The Two Brothers," in The Brothers Grimm, *The Complete Grimms' Fairy Tales* (translated by Margaret Hunt and James Stern; New York: Pantheon Books, 1944), pp. 308–11.

16. *The Mabinogion and Other Medieval Welsh Tales* (translated by Patrick K. Ford; Berkeley: University of California Press, 1977), pp. 37–56.

17. *Volsungasaga (The Saga of the Volsungs)* (translated by Jesse Byock; Berkeley: University of California Press, 1990), #29, "Sigurd Rides through the Wavering Flames of Brynhild," p. 80.

18. Susan Feldmann, "The Twin Brotheres," in *African Myths and Tales* (New York: Dell, 1963), pp. 272–6.

19. Stith Thompson, *Motif-Index of Folk Literature* (6 vols.; Bloomington, IN: Indiana University Press, 1955–58).

20. *Le Petit Robert 2,* p. 1690: Sosie, nom de l'esclave d'Amphitryon dont Mercure prend l'aspect. Personne qui a une parfaite ressemblance avec une autre. Paul Robert, *Le nouveau petit Robert, Dictionnaire alphabetique et analogique de la langue francaise.* Nouvelle edition. Paris: Dictionnaires Le Robert, 1996.

21. *Le Petit Robert 1,* p. 1838.

22. Plautus, *Amphitryon,* 1. 422, translated by James. H. Mantinband, in Charles E. Passage and James H. Mantinband, editors and translators, *Amphitryon: Three Plays*

in New Verse Translation (Plautus, Moliere, Kleist), Together with a Comprehensive Account of the Evolution of the Legend and Its Subsequent History on the Stage (Chapel Hill: University of North Carolina Press, 1974), p. 58.

23. Otto Rank, *The Double: A Psychoanalytic Study* (translated by Harry Tucker, Jr.; New York: New American Library, 1971 [1925]), p. 33.

24. Fyodor Dostoyevski, *The Double: Two Versions* (translated by Evelyn Harden; Ann Arbor, MI: Ardis, 1985 [*Dvoynik*, 1846]).

25. Wendy Doniger, "When a Kiss Is Still a Kiss: Memories of the Mind and the Body in Ancient India and Hollywood," in *Kenyon Review* (XIX: 1, Winter, 1997), pp. 118–33.

26. Jeanine Basinger, *How Hollywood Spoke to Women, 1930–1960* (New York: Alfred Knopf, 1993), p. 100.

27. "Sister Charged with Murder, and Identity Switch," *New York Times,* Thursday, July 17, 1997, p. A16.

28. Richard Winnington, cited by Leslie Halliwell (*Halliwell's Film Guide,* edited by John Walker; New York: Harper, 1995).

29. Wendy Doniger, "Myths and Methods in the Dark," in *Journal of Religion* (76:4, October, 1996), pp. 531–47.

30. Wendy Doniger, *Splitting Women in Indian and Greek Myths* (The 1996–7 Jordan Lectures at the University of London).

31. *Tristan,* of Gottfried von Strassburg. Translated by A. T. Hatto and supplemented with the surviving fragments of the *Tristan* of Thomas (Harmondsworth: Penguin Books, 1960).

32. Wendy Doniger O'Flaherty, *Dreams, Illusion, and Other Realities* (Chicago: University of Chicago Press, 1984), pp. 95–6.

33. Cathy Winkler, "Rape Trauma: Contexts of Meaning," in *Embodiment and Experience: The Existential Ground of Culture and Self* (edited by Thomas Csordas; Cambridge: Cambridge University Press, 1994), pp. 256–7.

34. Havelock Ellis, "The Psychic State in Pregnancy," in *Studies in the Psychology of Sex* (Philadelphia: F. A. Davis, 1906), vol. 5, pp. 201–29.

35. *Mahabharata* (Poona: Bhandarkar Oriental Research Institute, 1933–69) 1.99–100; Wendy Doniger O'Flaherty, *Textual Sources for the Study of Hinduism* (Chicago: University of Chicago Press, 1990), pp. 46–51.

36. Peter Steinfels, "Beliefs," *New York Times,* Saturday, July 12, 1997, p. 9.

37. See Chinua Achebe, *Things Fall Apart* (New York: Knopf/Random House, 1992).

PART III
Ethics and Religion

Cloning Human Beings: An Assessment of the Ethical Issues Pro and Con

Dan W. Brock

The world of science and the public at large were both shocked and fascinated by the announcement in the journal *Nature* by Ian Wilmut and his colleagues that they had successfully cloned a sheep from a single cell of an adult sheep (Wilmut, 1997). But many were troubled or apparently even horrified at the prospect that cloning of adult humans by the same process might be possible as well. The response of most scientific and political leaders to the prospect of human cloning, indeed of Dr. Wilmut as well, was of immediate and strong condemnation.

A few more cautious voices were heard both suggesting some possible benefits from the use of human cloning in limited circumstances and questioning its too quick prohibition, but they were a clear minority. A striking feature of these early responses was that their strength and intensity seemed far to outrun the arguments and reasons offered in support of them—they seemed often to be "gut level" emotional reactions rather than considered reflections on the issues. Such reactions should not be simply dismissed, both because they may point us to important considerations otherwise missed and not easily articulated, and because they often have a major impact on public policy. But the formation of public policy should not ignore the moral reasons and arguments that bear on the practice of human cloning—these must be articulated in order to understand and inform people's more immediate emotional responses. This essay is an effort to articulate, and to evaluate critically, the main moral considerations and arguments for and against human cloning. Though many people's religious beliefs inform their views on human cloning, and it is often difficult to separate religious from secular positions, I shall restrict myself to arguments and reasons that can be given a clear secular formulation.

On each side of the issue there are two distinct kinds of moral arguments brought forward. On the one hand, some opponents claim that human cloning would violate fundamental moral or human rights, while some proponents argue that its prohibition would violate such rights. While moral and even human rights need not be understood as absolute, they do place moral restrictions on permissible actions that an appeal to a mere balance of benefits over harms cannot justify overriding; for example, the rights of human subjects in research must be respected even if the result is that some potentially beneficial research is more difficult or cannot be done. On the other hand, both opponents and proponents also cite the likely harms and benefits, both to individuals and to society, of the practice. I shall begin with the arguments in support of permitting human cloning, although with no implication that it is the stronger or weaker position.

Moral Arguments in Support of Human Cloning

IS THERE A MORAL RIGHT TO USE HUMAN CLONING?

What moral right might protect at least some access to the use of human cloning? A commitment to individual liberty, such as defended by J. S. Mill, requires that individuals be left free to use human cloning if they so choose and if their doing so does not cause significant harms to others, but liberty is too broad in scope to be an uncontroversial moral right (Mill, 1859; Rhodes, 1995). Human cloning is a means of reproduction (in the most literal sense) and so the most plausible moral right at stake in its use is a right to reproductive freedom or procreative liberty (Robertson, 1994a; Brock, 1994), understood to include both the choice not to reproduce, for example, by means of contraception or abortion, and also the right to reproduce.

The right to reproductive freedom is properly understood to include the right to use various assisted reproductive technologies (ARTs), such as in vitro fertilization (IVF), oocyte donation, and so forth. The reproductive right relevant to human cloning is a negative right, that is, a right to use ARTs without interference by the government or others when made available by a willing provider. The choice of an assisted means of reproduction should be protected by reproductive freedom even when it is not the only means for individuals to reproduce, just as the choice among different means of preventing conception is protected by reproductive freedom. However, the case for permitting the use of a particular means of reproduction is strongest when it is necessary for particular individuals to be able to procreate at all, or to do so without great burdens or harms to themselves or others. In some cases human cloning could be the only means for individuals to procreate while retaining a biological tie to their child, but in other cases different means of procreating might also be possible.

It could be argued that human cloning is not covered by the right to reproductive freedom because whereas current ARTs and practices covered by that right are remedies for inabilities to reproduce sexually, human cloning is an entirely new means of reproduction; indeed, its critics see it as more a means of manufacturing humans than of reproduction. Human cloning is a different means of reproduction than sexual reproduction, but it is a means that can serve individuals' interest in reproducing. If it is not protected by the moral right to reproductive freedom, I believe that must be not because it is a new means of reproducing, but instead because it has other objectionable or harmful features; I shall evaluate these other ethical objections to it later.

When individuals have alternative means of procreating, human cloning typically would be chosen because it replicates a particular individual's genome. The reproductive interest in question then is not simply reproduction itself, but a more specific interest in choosing what kind of children to have. The right to reproductive freedom is usually understood to cover at least some choice about the kind of children one will have. Some individuals choose reproductive partners in the hope of producing offspring with desirable traits. Genetic testing of fetuses or preimplantation embryos for genetic disease or abnormality is done to avoid having a child with those diseases or abnormalities. Respect for individual self-determination, which is one of the grounds of a moral right to reproductive freedom, includes respecting individuals' choices about whether to have a child with a condition that will place severe burdens on them, and cause severe burdens to the child itself.

The less a reproductive choice is primarily the determination of one's own life, but primarily the determination of the nature of another, as in the case of human cloning, the more moral weight the interests of that other person, that is the cloned child, should have in decisions that determine its nature (Annas, 1994). But even then parents are typically accorded substantial, but not unlimited, discretion in shaping the persons their children will become, for example, through education and other childrearing decisions. Even if not part of reproductive freedom, the right to raise one's children as one sees fit, within limits mostly determined by the interests of the children, is also a right to determine within limits what kinds of persons one's children will become. This right includes not just preventing certain diseases or harms to children, but selecting and shaping desirable features and traits in one's children. The use of human cloning is one way to exercise that right.

Public policy and the law now permit prospective parents to conceive, or to carry a conception to term, when there is a significant risk or even certainty that the child will suffer from a serious genetic disease. Even when others think the risk or certainty of genetic disease makes it morally wrong to conceive, or to carry a fetus to term, the parents' right to reproductive freedom permits them to do so. Most possible harms to a cloned child are less serious than the genetic harms with which parents can now permit their offspring to be conceived or born.

I conclude that there is good reason to accept that a right to reproductive freedom presumptively includes both a right to select the means of reproduction, as well as a right to determine what kind of children to have, by use of human cloning. However, the specific reproductive interest of determining what kind of children to have is less weighty than are other reproductive interests and choices whose impact falls more directly and exclusively on the parents rather than the child. Even if a moral right to reproductive freedom protects the use of human cloning, that does not settle the moral issue about human cloning, since there may be other moral rights in conflict with this right, or serious enough harms from human cloning to override the right to use it; this right can be thought of as establishing a serious moral presumption supporting access to human cloning.

WHAT INDIVIDUAL OR SOCIAL BENEFITS MIGHT HUMAN CLONING PRODUCE?

LARGELY INDIVIDUAL BENEFITS

The literature on human cloning by nuclear transfer or by embryo splitting contains a few examples of circumstances in which individuals might have good reasons to want to use human cloning. However, human cloning seems not to be the unique answer to any great or pressing human need and its benefits appear to be limited at most. What are the principal possible benefits of human cloning that might give individuals good reasons to want to use it?

1. *Human cloning would be a new means to relieve the infertility some persons now experience.* Human cloning would allow women who have no ova or men who have no sperm to produce an offspring that is biologically related to them (Eisenberg, 1976; Robertson, 1994b, 1997; LaBar, 1984). Embryos might also be cloned, by either nuclear transfer or embryo splitting, in order to increase the number of embryos for implantation and improve the chances of successful conception (NABER, 1994). The benefits from human cloning to relieve infertility are greater the more persons there are who cannot overcome their infertility by any other means acceptable to them. I do not know of data on this point, but the numbers who would use cloning for this reason are probably not large.

The large number of children throughout the world possibly available for adoption represents an alternative solution to infertility only if we are prepared to discount as illegitimate the strong desire of many persons, fertile and infertile, for the experience of pregnancy and for having and raising a child biologically related to them. While not important to all infertile (or fertile) individuals, it is important to many and is respected and met through other forms of assisted reproduction that maintain a biological connection when that is possible; that desire does not become illegitimate simply because human cloning would be the best or only means of overcoming an individual's infertility.

2. *Human cloning would enable couples in which one party risks transmitting a serious hereditary disease to an offspring to reproduce without doing so* (Robertson, 1994b). By using donor sperm or egg donation, such hereditary risks can generally be avoided now without the use of human cloning. These procedures may be unacceptable to some couples, however, or at least considered less desirable than human cloning because they introduce a third party's genes into their reproduction instead of giving their offspring only the genes of one of them. Thus, in some cases human cloning could be a rea-

sonable means of preventing genetically transmitted harms to off-spring. Here too, we do not know how many persons would want to use human cloning instead of other means of avoiding the risk of genetic transmission of a disease or of accepting the risk of transmitting the disease, but the numbers again are probably not large.

3. *Human cloning to make a later twin would enable a person to obtain needed organs or tissues for transplantation* (Robertson, 1994b, 1997; Kahn, 1989; Harris, 1992). Human cloning would solve the problem of finding a transplant donor whose organ or tissue is an acceptable match and would eliminate, or drastically reduce, the risk of transplant rejection by the host. The availability of human cloning for this purpose would amount to a form of insurance to enable treatment of certain kinds of medical conditions. Of course, sometimes the medical need would be too urgent to permit waiting for the cloning, gestation, and development that is necessary before tissues or organs can be obtained for transplantation. In other cases, taking an organ also needed by the later twin, such as a heart or a liver, would be impermissible because it would violate the later twin's rights.

Such a practice can be criticized on the ground that it treats the later twin not as a person valued and loved for his or her own sake, as an end in itself in Kantian terms, but simply as a means for benefiting another. This criticism assumes, however, that only this one motive defines the reproduction and the relation of the person to his or her later twin. The well-known case some years ago in California of the Ayalas, who conceived in the hopes of obtaining a source for a bone marrow transplant for their teenage daughter suffering from leukemia, illustrates the mistake in this assumption. They argued that whether or not the child they conceived turned out to be a possible donor for their daughter, they would value and love the child for itself, and treat it as they would treat any other member of their family. That one reason they wanted it, as a possible means to saving their daughter's life, did not preclude their also loving and valuing it for its own sake; in Kantian terms, it was treated as a possible means to saving their daughter, but not *solely as a means,* which is what the Kantian view proscribes.

Indeed, when people have children, whether by sexual means or with the aid of ARTs, their motives and reasons for doing so are typically many and complex, and include reasons less laudable than obtaining lifesaving medical treatment, such as having someone who needs them, enabling them to live on their own, qualifying for government benefit programs, and so forth. While these are not admirable motives for having children and may not bode well for the child's upbringing and future, public policy does not assess prospective parents' motives and reasons for procreating as a condition of their doing so.

4. *Human cloning would enable individuals to clone someone who had special meaning to them, such as a child who had died* (Robertson, 1994b). There is no denying that if human cloning were available, some individuals would want to use it for this purpose, but their desire usually would be based on a deep confusion. Cloning such a child would not replace the child the parents had loved and lost, but would only create a different child with the same genes. The child they loved and lost was a unique individual who had been shaped by his or her environment and choices, not just his or her genes, and more importantly who had experienced a particular relationship with them. Even if the later cloned child could not only have the same genes but also be subjected to the same environment, which of course is impossible, it would remain a different child than the one they had loved and lost because it would share a different history with them (Thomas, 1974). Cloning the lost child might help the parents accept and move on from their loss, but another already existing sibling or a new child that was not a clone might do this equally well; indeed, it might do so better since the appearance of the cloned later twin would be a constant reminder of the child they had lost. Nevertheless, if human cloning enabled some individuals to clone a person who had special meaning to

them and doing so gave them deep satisfaction, that would be a benefit to them even if their reasons for wanting to do so, and the satisfaction they in turn received, were based on a confusion.

LARGELY SOCIAL BENEFITS

5. *Human cloning would enable the duplication of individuals of great talent, genius, character, or other exemplary qualities.* Unlike the first four reasons for human cloning which appeal to benefits to specific individuals, this reason looks to benefits to the broader society from being able to replicate extraordinary individuals—a Mozart, Einstein, Gandhi, or Schweitzer (Lederberg, 1966; McKinnell, 1979). Much of the appeal of this reason, like much support and opposition to human cloning, rests largely on a confused and false assumption of genetic determinism, that is, that one's genes fully determine what one will become, do, and accomplish. What made Mozart, Einstein, Gandhi, and Schweitzer the extraordinary individuals they were was the confluence of their particular genetic endowments with the environments in which they were raised and lived and the particular historical moments they in different ways seized. Cloning them would produce individuals with the same genetic inheritances (nuclear transfer does not even produce 100 percent genetic identity, although for the sake of exploring the moral issues I have followed the common assumption that it does), but it is not possible to replicate their environments or the historical contexts in which they lived and their greatness flourished. We do not know the degree or specific respects in which any individual's greatness depended on "nature" or "nurture," but we do know that it always depends on an interaction of them both. Cloning could not even replicate individuals' extraordinary capabilities, much less their accomplishments, because these too are the product of their inherited genes and their environments, not of their genes alone.

None of this is to deny that Mozart's and Einstein's extraordinary musical and intellectual capabilities, nor even Gandhi's and

Schweitzer's extraordinary moral greatness, were produced in part by their unique genetic inheritances. Cloning them might well produce individuals with exceptional capacities, but we simply do not know how close their clones would be in capacities or accomplishments to the great individuals from whom they were cloned. Even so, the hope for exceptional, even if less and different, accomplishment from cloning such extraordinary individuals might be a reasonable ground for doing so.

Worries here about abuse, however, surface quickly. Whose standards of greatness would be used to select individuals to be cloned? Who would control use of human cloning technology for the benefit of society or mankind at large? Particular groups, segments of society, or governments might use the technology for their own benefit, under the cover of benefiting society or even mankind at large.

6. *Human cloning and research on human cloning might make possible important advances in scientific knowledge, for example, about human development* (Walters, 1982; Smith, 1983). While important potential advances in scientific or medical knowledge from human cloning or human cloning research have frequently been cited, there are at least three reasons for caution about such claims. First, there is always considerable uncertainty about the nature and importance of the new scientific or medical knowledge to which a dramatic new technology like human cloning will lead; the road to new knowledge is never mapped in advance and takes many unexpected turns. Second, we do not know what new knowledge from human cloning or human cloning research could also be gained by other means that do not have the problematic moral features to which its opponents object. Third, what human cloning research would be compatible with ethical and legal requirements for the use of human subjects in research is complex, controversial, and largely unexplored. Creating human clones solely for the purpose of research would be to use them solely for the benefit of others without their consent, and so unethical. But if and when human cloning was established to be

safe and effective, then new scientific knowledge might be obtained from its use for legitimate, nonresearch reasons.

Although there is considerable uncertainty concerning most of human cloning's possible individual and social benefits that I have discussed, and although no doubt it could have other benefits or uses that we cannot yet envisage, I believe it is reasonable to conclude at this time that human cloning does not seem to promise great benefits or uniquely to meet great human needs. Nevertheless, despite these limited benefits, a moral case can be made that freedom to use human cloning is protected by the important moral right to reproductive freedom. I shall turn now to what moral rights might be violated, or harms produced, by research on or use of human cloning.

Moral Arguments Against Human Cloning

WOULD THE USE OF HUMAN CLONING VIOLATE IMPORTANT MORAL RIGHTS?

Many of the immediate condemnations of any possible human cloning following Wilmut's cloning of Dolly claimed that it would violate moral or human rights, but it was usually not specified precisely, or often even at all, what rights would be violated (WHO, 1997). I shall consider two possible candidates for such a right: a right to have a unique identity and a right to ignorance about one's future or to an open future. Claims that cloning denies individuals a unique identity are common, but I shall argue that even if there is a right to a unique identity, it could not be violated by human cloning. The right to ignorance or to an open future has only been explicitly defended, to my knowledge, by two commentators, and in the context of human cloning, only by Hans Jonas; it supports a more promising, but in my view ultimately unsuccessful, argument that human cloning would violate an important moral or human right.

Is there a moral or human right to a unique identity, and if so would it be violated by human cloning? For human cloning to violate a right to a unique identity, the relevant sense of identity would have to be genetic identity, that is, a right to a unique unrepeated genome. This would be violated by human cloning, but is there any such right? It might be thought that cases of identical twins show there is no such right because no one claims that the moral or human rights of the twins have been violated. However, this consideration is not conclusive (Kass, 1985; NABER, 1994). Only human actions can violate others' rights; outcomes that would constitute a rights violation if deliberately caused by human action are not a rights violation if a result of natural causes. If Arthur deliberately strikes Barry on the head so hard as to cause his death, he violates Barry's right not to be killed; if lightning strikes Cheryl, causing her death, her right not to be killed has not been violated. Thus, the case of twins does not show that there could not be a right to a unique genetic identity.

What is the sense of identity that might plausibly be what each person has a right to have uniquely, that constitutes the special uniqueness of each individual (Macklin 1994; Chadwick 1982)? Even with the same genes, homozygous twins are numerically distinct and not identical, so what is intended must be the various properties and characteristics that make each individual qualitatively unique and different from others. Does having the same genome as another person undermine that unique qualitative identity? Only on the crudest genetic determinism, according to which an individual's genes completely and decisively determine everything else about the individual, all his or her other nongenetic features and properties, together with the entire history or biography that constitutes his or her life. But there is no reason whatever to believe that kind of genetic determinism. Even with the same genes, differences in genetically identical twins' psychological and personal characteristics develop over time together with differences in their life histories, personal relationships, and life choices;

sharing an identical genome does not prevent twins from developing distinct and unique personal identities of their own.

We need not pursue whether there is a moral or human right to
a unique identity—no such right is found among typical accounts
and enumerations of moral or human rights—because even if there
is such a right, sharing a genome with another individual as a result
of human cloning would not violate it. The idea of the uniqueness,
or unique identity, of each person historically predates the development of modern genetics. A unique genome thus could not be
the ground of this long-standing belief in the unique human identity of each person.

I turn now to whether human cloning would violate what Hans
Jonas called a right to ignorance, or what Joel Feinberg called a
right to an open future (Jonas, 1974; Feinberg, 1980). Jonas argued
that human cloning in which there is a substantial time gap between the beginning of the lives of the earlier and later twin is fundamentally different from the simultaneous beginning of the lives
of homozygous twins that occur in nature. Although contempora-
neous twins begin their lives with the same genetic inheritance,
they do so at the same time, and so in ignorance of what the other
who shares the same genome will by his or her choices make of his
or her life.

A later twin created by human cloning, Jonas argues, knows, or
at least believes she knows, too much about herself. For there is already in the world another person, her earlier twin, who from the
same genetic starting point has made the life choices that are still
in the later twin's future. It will seem that her life has already been
lived and played out by another, that her fate is already determined;
she will lose the sense of human possibility in freely and spontaneously creating her own future and authentic self. It is tyrannical,
Jonas claims, for the earlier twin to try to determine another's fate
in this way.

Jonas's objection can be interpreted so as not to assume either
a false genetic determinism, or a belief in it. A later twin might

grant that he is not determined to follow in his earlier twin's foot-
steps, but nevertheless the earlier twin's life might always haunt
him, standing as an undue influence on his life, and shaping it in
ways to which others' lives are not vulnerable. But the force of the
objection still seems to rest on the false assumption that having the
same genome as his earlier twin unduly restricts his freedom to create a different life and self than the earlier twin's. Moreover, a family environment also importantly shapes children's development,
but there is no force to the claim of a younger sibling that the existence of an older sibling raised in that same family is an undue influence on the younger sibling's freedom to make his own life for
himself in that environment. Indeed, the younger twin or sibling
might gain the benefit of being able to learn from the older twin's
or sibling's mistakes.

A closely related argument can be derived from what Joel Feinberg has called a child's right to an open future. This requires that
others raising a child not so close off the future possibilities that the
child would otherwise have as to eliminate a reasonable range of
opportunities for the child autonomously to construct his or her
own life. One way this right might be violated is to create a later
twin who will believe her future has already been set for her by the
choices made and the life lived by her earlier twin.

The central difficulty in these appeals to a right either to ignorance or to an open future is that the right is not violated merely
because the later twin is likely to *believe* that his future is already determined, when that belief is clearly false and supported only by the
crudest genetic determinism. If we know the later twin will falsely
believe that his open future has been taken from him as a result of
being cloned, even though in reality it has not, then we know that
cloning will cause the twin psychological distress, but not that it
will violate his right. Jonas's right to ignorance, and Feinberg's
right of a child to an open future, are not not violated by human
cloning, though they do point to psychological harms that a later
twin may be likely to experience and that I will take up later.

Neither a moral or human right to a unique identity, nor one to ignorance and an open future, would be violated by human cloning. There may be other moral or human rights that human cloning would violate, but I do not know what they might be. I turn now to consideration of the harms that human cloning might produce.

WHAT INDIVIDUAL OR SOCIAL HARMS MIGHT HUMAN CLONING PRODUCE?

There are many possible individual or social harms that have been posited by one or another commentator and I shall only try to cover the more plausible and significant of them.

LARGELY INDIVIDUAL HARMS

1. Human cloning would produce psychological distress and harm in the later twin. No doubt knowing the path in life taken by one's earlier twin might often have several bad psychological effects (Callahan, 1993; LaBar, 1984; Macklin, 1994; McCormick, 1993; Studdard, 1978; Rainer, 1978; Verhey, 1994). The later twin might feel, even if mistakenly, that her fate has already been substantially laid out, and so have difficulty freely and spontaneously taking responsibility for and making her own fate and life. The later twin's experience or sense of autonomy and freedom might be substantially diminished, even if in actual fact they are diminished much less than it seems to her. She might have a diminished sense of her own uniqueness and individuality, even if once again these are in fact diminished little or not at all by having an earlier twin with the same genome. If the later twin is the clone of a particularly exemplary individual, perhaps with some special capabilities and accomplishments, she might experience excessive pressure to reach the very high standards of ability and accomplishment of the earlier twin (Rainer, 1978). These various psychological effects might take a heavy toll on the later twin and be serious burdens to her.

While psychological harms of these kinds from human cloning are certainly possible, and perhaps even likely in some cases, they remain at this point only speculative since we have no experience with human cloning and the creation of earlier and later twins. Nevertheless, if experience with human cloning confirmed that serious and unavoidable psychological harms typically occurred to the later twin, that would be a serious moral reason to avoid the practice. Intuitively at least, psychological burdens and harms seem more likely and more serious for a person who is only one of many identical later twins cloned from one original source, so that the clone might run into another identical twin around every street corner. This prospect could be a good reason to place sharp limits on the number of twins that could be cloned from any one source.

One argument has been used by several commentators to undermine the apparent significance of potential psychological harms to a later twin (Chadwick, 1982; Robertson, 1994b, 1997; Macklin, 1994). The point derives from a general problem, called the nonidentity problem, posed by the philosopher Derek Parfit, although not originally directed to human cloning (Parfit, 1984). Here is the argument. Even if all these psychological burdens from human cloning could not be avoided for any later twin, they are not harms to the twin, and so not reasons not to clone the twin. That is because the only way for the twin to avoid the harms is never to be cloned, and so never to exist at all. But these psychological burdens, hard though they might be, are not so bad as to make the twin's life, all things considered, not worth living. So the later twin is not harmed by being given a life even with these psychological burdens, since the alternative of never existing at all is arguably worse—he or she never has a worthwhile life—but certainly not better for the twin. And if the later twin is not harmed by having been created with these unavoidable burdens, then how could he or she be wronged by having been created with them? And if the later twin is not wronged, then why is any wrong being done by human cloning? This argument has considerable potential import, for if it is sound it will undermine the apparent moral importance of any bad consequence of human cloning to the later twin that is not so serious as to make the twin's life, all things considered, not worth living.

I defended elsewhere the position regarding the general case of genetically transmitted handicaps, that if one could have a *different* child without comparable burdens (for the case of cloning, by using a different method of reproduction which did not result in a later twin), there is as strong a moral reason to do so as there would be not to cause similar burdens to an already existing child (Brock, 1995). Choosing to create the later twin with serious psychological burdens instead of a different person who would be free of them, without weighty overriding reasons for choosing the former, would be morally irresponsible or wrong, even if doing so does not harm or wrong the later twin who could only exist with the burdens. These issues are too detailed and complex to pursue here and the nonidentity problem remains controversial and not fully resolved, but at the least, the argument for disregarding the psychological burdens to the later twin because he or she could not exist without them is controversial, and in my view mistaken. Such psychological harms, as I shall continue to call them, are speculative, but they should not be disregarded because of the nonidentity problem.

2. Human cloning procedures would carry unacceptable risks to the clone. There is no doubt that attempts to clone a human being at the present time would carry unacceptable risks to the clone. Further research on the procedure with animals, as well as research to establish its safety and effectiveness for humans, is clearly necessary before it would be ethical to use the procedure on humans. One risk to the clone is the failure to implant, grow, and develop successfully, but this would involve the embryo's death or destruction long before most people or the law consider it to be a person with moral or legal protections of its life.

Other risks to the clone are that the procedure in some way goes wrong, or unanticipated harms come to the clone; for example, Harold Varmus, director of the National Institutes of Health, raised the concern that a cell many years old from which a person is cloned could have accumulated genetic mutations during its years in another adult that could give the resulting clone a predisposition to cancer or other diseases of aging (Weiss, 1997). Risks to an ovum donor (if any), a nucleus donor, and a woman who receives the embryo for implantation would likely be ethically acceptable with the informed consent of the involved parties.

I believe it is too soon to say whether unavoidable risks to the clone would make human cloning forever unethical. At a minimum, further research is needed to better define the potential risks to humans. But we should not insist on a standard that requires risks to be lower than those we accept in sexual reproduction, or in other forms of ART.

LARGELY SOCIAL HARMS

3. *Human cloning would lessen the worth of individuals and diminish respect for human life.* Unelaborated claims to this effect were common in the media after the announcement of the cloning of Dolly. Ruth Macklin explored and criticized the claim that human cloning would diminish the value we place on, and our respect for, human life because it would lead to persons being viewed as replaceable (Macklin, 1994). As I have argued concerning a right to a unique identity, only on a confused and indefensible notion of human identity is a person's identity determined solely by his or her genes, and so no individual could be fully replaced by a later clone possessing the same genes. Ordinary people recognize this clearly. For example, parents of a child dying of a fatal disease would find it insensitive and ludicrous to be told they should not grieve for their coming loss because it is possible to replace him by cloning him; it is *their child who is dying* whom they love and value, and that child and his importance to them is not replaceable by a cloned later twin. Even if they would also come to love and value a later twin as much as they now love and value their child who is dying, that would be to love and value that *different child* for its own sake, not as a replace-

ment for the child they lost. Our relations of love and friendship are with distinct, historically situated individuals with whom over time we have shared experiences and our lives, and whose loss to us can never be replaced.

A different version of this worry is that human cloning would result in persons' worth or value seeming diminished because we would come to see persons as able to be manufactured or "handmade." This demystification of the creation of human life would reduce our appreciation and awe of human life and of its natural creation. It would be a mistake, however, to conclude that a person created by human cloning is of less value or is less worthy of respect than one created by sexual reproduction. At least outside of some religious contexts, it is the nature of a being, not how it is created, that is the source of its value and makes it worthy of respect. For many people, gaining a scientific understanding of the truly extraordinary complexity of human reproduction and development increases, instead of decreases, their awe of the process and its product.

A more subtle route by which the value we place on each individual human life might be diminished could come from the use of human cloning with the aim of creating a child with a particular genome, either the genome of another individual especially meaningful to those doing the cloning or an individual with exceptional talents, abilities, and accomplishments. The child then comes to be objectified, valued only as an object and for its genome, or at least for its genome's expected phenotypic expression, and no longer recognized as having the intrinsic equal moral value of all persons, simply as persons. For the moral value and respect due all persons to come to be seen as resting only on the instrumental value of individuals and of their particular qualities to others would be to fundamentally change the moral status properly accorded to persons. Individuals would lose their moral standing as full and equal members of the moral community, replaced by the different instrumental value each has to others.

Such a change in the equal moral value and worth accorded to persons should be avoided at all costs, but it is far from clear that such a change would result from permitting human cloning. Parents, for example, are quite capable of distinguishing their children's intrinsic value, just as individual persons, from their instrumental value based on their particular qualities or properties. The equal moral value and respect due all persons simply as persons is not incompatible with the different instrumental value of different individuals; Einstein and an untalented physics graduate student have vastly different value as scientists, but share and are entitled to equal moral value and respect as persons. It is a confused mistake to conflate these two kinds of value and respect. If making a large number of clones from one original person would be more likely to foster it, that would be a further reason to limit the number of clones that could be made from one individual.

4. *Human cloning might be used by commercial interests for financial gain.* Both opponents and proponents of human cloning agree that cloned embryos should not be able to be bought and sold. In a science fiction frame of mind, one can imagine commercial interests offering genetically certified and guaranteed embryos for sale, perhaps offering a catalogue of different embryos cloned from individuals with a variety of talents, capacities, and other desirable properties. This would be a fundamental violation of the equal moral respect and dignity owed to all persons, treating them instead as objects to be differentially valued, bought, and sold in the marketplace. Even if embryos are not yet persons at the time they would be purchased or sold, they would be being valued, bought, and sold for the persons they will become. The moral consensus against any commercial market in embryos, cloned or otherwise, should be enforced by law whatever the public policy ultimately is on human cloning.

5. *Human cloning might be used by governments or other groups for immoral and exploitative purposes.* In *Brave New World,* Aldous Huxley imagined cloning individuals who have been engineered with lim-

ited abilities and conditioned to do, and to be happy doing, the menial work that society needed done (Huxley, 1932). Selection and control in the creation of people was exercised not in the interests of the persons created, but in the interests of the society and at the expense of the persons created; nor did it serve individuals' interests in reproduction and parenting. Any use of human cloning for such purposes would exploit the clones solely as means for the benefit of others, and would violate the equal moral respect and dignity they are owed as full moral persons. If human cloning is permitted to go forward, it should be with regulations that would clearly prohibit such immoral exploitation.

Fiction contains even more disturbing or bizarre uses of human cloning, such as Mengele's creation of many clones of Hitler in Ira Levin's *The Boys from Brazil* (Levin, 1976), Woody Allen's science fiction cinematic spoof *Sleeper* in which a dictator's only remaining part, his nose, must be destroyed to keep it from being cloned, and the contemporary science fiction film *Blade Runner*. These nightmare scenarios may be quite improbable, but their impact should not be underestimated on public concern with technologies like human cloning. Regulation of human cloning must assure the public that even such far-fetched abuses will not take place.

Conclusion

Human cloning has until now received little serious and careful ethical attention because it was typically dismissed as science fiction, and it stirs deep, but difficult to articulate, uneasiness and even revulsion in many people. Any ethical assessment of human cloning at this point must be tentative and provisional. Fortunately, the science and technology of human cloning are not yet in hand, and so a public and professional debate is possible without the need for a hasty, precipitate policy response.

The ethical pros and cons of human cloning, as I see them at this time, are sufficiently balanced and uncertain that there is not an ethically decisive case either for or against permitting it or doing it. Access to human cloning can plausibly be brought within a moral right to reproductive freedom, but its potential legitimate uses appear few and do not promise substantial benefits. It is not a central component of the moral right to reproductive freedom and it does not uniquely serve any major or pressing individual or social needs. On the other hand, contrary to the pronouncements of many of its opponents, human cloning seems not to be a violation of moral or human rights. But it does risk some significant individual or social harms, although most are based on common public confusions about genetic determinism, human identity, and the effects of human cloning. Because most potential harms feared from human cloning remain speculative, they seem insufficient to warrant at this time a complete legal prohibition of either research on or later use of human cloning, if and when its safety and efficacy are established. Legitimate moral concerns about the use and effects of human cloning, however, underline the need for careful public oversight of research on its development, together with a wider public and professional debate and review before cloning is used on human beings.

References

Annas, G. J. (1994). "Regulatory Models for Human Embryo Cloning: The Free Market, Professional Guidelines, and Government Restrictions." *Kennedy Institute of Ethics Journal* 4,3:235–249.

Brock, D. W. (1994). "Reproductive Freedom: Its Nature, Bases and Limits," in *Health Care Ethics: Critical Issues for Health Professionals,* eds. D. Thomasma and J. Monagle. Gaithersbrug, MD: Aspen Publishers.

Brock, D. W. (1995). "The Non-Identity Problem and Genetic Harm." *Bioethics* 9:269–275.

Callahan, D. (1993). "Perspective on Cloning: A Threat to Individual Uniqueness." *Los Angeles Times,* November 12, 1993:B7.

Chadwick, R. F. (1982). "Cloning." *Philosophy* 57:201–209.

Eisenberg, L. (1976). "The Outcome as Cause: Predestination and Human Cloning." *The Journal of Medicine and Philosophy* 1:318–331.

Feinberg, J. (1980). "The Child's Right to an Open Future," in *Whose Child? Children's Rights, Parental Authority, and State Power,* eds. W. Aiken and H. LaFollette. Totowa, NJ: Rowman and Littlefield.

Harris, J. (1992). *Wonderwoman and Superman: The Ethics of Biotechnology.* Oxford: Oxford University Press.

Huxley, A. (1932). *Brave New World.* London: Chalto and Winders.

Jonas, H. (1974). *Philosophical Essays: From Ancient Creed to Technological Man.* Englewood Cliffs, NJ: Prentice-Hall.

Kahn, C. (1989). "Can We Achieve Immortality?" *Free Inquiry* 9:14–18.

Kass, L. (1985). *Toward a More Natural Science.* New York: The Free Press.

LaBar, M. (1984). "The Pros and Cons of Human Cloning." *Thought* 57:318–333.

Lederberg, J. (1966). "Experimental Genetics and Human Evolution." *The American Naturalist* 100:519–531.

Levin, I. (1976). *The Boys from Brazil.* New York: Random House.

Macklin, R. (1994). "Splitting Embryos on the Slippery Slope: Ethics and Public Policy." *Kennedy Institute of Ethics Journal* 4:209–226.

McCormick, R. (1993). "Should We Clone Humans?" *Christian Century* 110:1148–1149.

McKinnell, R. (1979). *Cloning: A Biologist Reports.* Minneapolis, MN: University of Minnesota Press.

Mill, J. S. (1859). *On Liberty.* Indianapolis, IN: Bobbs-Merrill Publishing.

NABER (National Advisory Board on Ethics in Reproduction) (1994). "Report on Human Cloning Through Embryo Splitting: An Amber Light." *Kennedy Institute of Ethics Journal* 4:251–282.

Parfit, D. (1984). *Reasons and Persons.* Oxford: Oxford University Press.

Rainer, J. D. (1978). "Commentary." *Man and Medicine: The Journal of Values and Ethics in Health Care* 3:115–117.

Rhodes, R. (1995). "Clones, Harms, and Rights." *Cambridge Quarterly of Healthcare Ethics* 4:285–290.

Robertson, J. A. (1994a). *Children of Choice: Freedom and the New Reproductive Technologies.* Princeton, NJ: Princeton University Press.

Robertson, J. A. (1994b). "The Question of Human Cloning." *Hastings Center Report* 24:6–14.

Robertson, J. A. (1997). "A Ban on Cloning and Cloning Research is Unjustified." Testimony Presented to the National Bioethics Advisory Commission, March 1997.

Smith, G. P. (1983). "Intimations of Immortality: Clones, Cyrons and the Law." *University of New South Wales Law Journal* 6:119–132.

Studdard, A. (1978). "The Lone Clone." *Man and Medicine: The Journal of Values and Ethics in Health Care* 3:109–114.

Thomas, L. (1974). "Notes of a Biology Watcher: On Cloning a Human Being." *New England Journal of Medicine* 291:1296–1297.

Verhey, A. D. (1994). "Cloning: Revisiting an Old Debate." *Kennedy Institute of Ethics Journal* 4:227–234.

Walters, W. A. W. (1982). "Cloning, Ectogenesis, and Hybrids: Things to Come?" in *Test-Tube Babies,* eds. W. A. W. Walters and P. Singer. Melbourne: Oxford University Press.

Weiss, R. (1997). "Cloning Suddenly Has Government's Attention." *International Herald Tribune,* March 7, 1997.

WHO (World Health Organization Press Office). (March 11, 1997). "WHO Director General Condemns Human Cloning." World Health Organization, Geneva, Switzerland.

Wilmut, I., et al. (1997). "Viable Offspring Derived from Fetal and Adult Mammalian Cells." *Nature* 385:810–813.

This essay is a shorter version of a paper prepared for the National Bioethics Advisory Commission.

I want to acknowledge with gratitude the invaluable help of my research assistant, Insoo Hyun, on this paper. He not only made it possible to complete the paper on the National Bioethics Advisory Commission's tight schedule, but also improved it with a number of insightful substantive suggestions.

Religious Perspectives

National Bioethics Advisory Commission

Religion and Human Cloning: An Historical Overview

I t is possible to identify four recent overlapping periods in which theologians and other religious thinkers have considered the scientific prospects and ethics of the cloning of humans. The first phase, which began in the mid-1960s and continued into the early 1970s, was shaped by a context of expanded choices and control of reproduction (e.g., the availability of the birth control pill), the prospects of alternative, technologically-assisted reproduction (e.g., *in vitro* fertilization [IVF]), and the advocacy by some biologists and geneticists of cloning "preferred" genotypes, which, in their view, would avoid overloading the human gene pool with genes that are linked to deleterious outcomes and that could place the survival of the human species at risk.

Several prominent theologians engaged in these initial discussions of human genetic manipulation and cloning, including Charles Curran, Bernard Häring, Richard McCormick, and Karl Rahner within Roman Catholicism, and Joseph Fletcher and Paul Ramsey within Protestantism. The diametrically opposed positions staked out by the last two theologians gave an early signal of the wide range of views that are still expressed by religious thinkers.

Joseph Fletcher advocated expansion of human freedom and control over human reproduction. He portrayed the cloning of humans as one of many present and prospective reproductive options that could be ethically justified by societal benefit. Indeed, for Fletcher, as a method of reproduction, cloning was preferable to the "genetic roulette" of sexual reproduction. He viewed laboratory reproduction as "radically human" because it is deliberate, designed, chosen, and willed (Fletcher, 1971, 1972, 1974, 1979).

By contrast, Paul Ramsey portrayed the cloning of humans as a "borderline" or moral boundary that could be crossed only at risk of compromise to humanity and to basic concepts of human procreation. Cloning threatened three "horizontal" (person-person) and two "vertical" (person-God) border crossings. First, clonal reproduction would require directed or managed breeding to serve the scientific ends of a controlled gene pool. Second, it would involve nontherapeutic experimentation on the unborn. Third, it would assault the meaning of parenthood by transforming "procreation" into "reproduction" and by severing the unitive end (expressing and sustaining mutual love) and the procreative end of human sexual expression. Fourth, the cloning of humans would express the sin of pride or hubris. Fifth, it could also be considered a sin of self-creation as humans aspire to become a "man-God" (Ramsey, 1966, 1970).

A second era of theological reflection on cloning humans began in 1978, a year that was notable for two events, the birth in Britain of the first IVF baby, Louise Brown, and the publication of David Rorvik's *In His Image,* an account alleging (falsely) the creation of the first cloned human being (Rorvik, 1978).

This period also witnessed the beginning of formal ecclesiastical involvement with questions of genetic manipulation. In 1977 the United Church of Christ produced a study booklet on *Genetic Manipulation,* which appears to be the earliest reference to human cloning among Protestant denominational literature (Lynn, 1977). It provided a general overview of the science and ethics of cloning humans but stopped short of a specific theological verdict.

The discussions of the 1970s continued into the 1980s with par-

ticular attention to IVF, artificial insemination by donor, and surrogacy. These techniques challenged traditional notions of the family by separating genetic and rearing fatherhood and genetic,
gestational, and rearing motherhood, as well as raising questions
about whether the contractual and commercial ties in many of
these arrangements were inimical to traditional religious views of
the family.

A third era of religious discussion began in 1993 with the report
from George Washington University of the separation of cells in
human blastomeres to create multiple, genetically identical embryos. The Roman Catholic Church expressed vigorous opposition to the procedure, and a Vatican editorial denounced the
research as "intrinsically perverse." Catholic moral theologians
invoked norms of individuality, dignity, and wholeness in condemning this research (McCormick, 1993, 1994). While many
Conservative Protestant scholars held that this research contra-
vened basic notions of personhood such as freedom, the sanctity of
life, and the image of God, some other Protestant scholars noted
its potential medical benefits and advocated careful regulation
rather than prohibition.

The fourth and most recent stage of religious discussion has
come in the wake of the successful cloning of Dolly the sheep
through the somatic cell nuclear transfer technique, as the cloning
of a human once again appeared to be a near-term possibility. Several Roman Catholic and Protestant thinkers have reiterated and re-
inforced past opposition and warnings.

However, some Protestant thinkers, in reflecting on the meaning of human partnership with ongoing divine creative activity,
have expressed qualified support for cloning research and for creating children using somatic cell nuclear transfer techniques. Likewise, some Jewish and Islamic thinkers encourage continuing
laboratory research on animal models and even laboratory work on
the possibility of cloning human beings (only in pursuit of a worthy objective), while expressing deep moral reservations, at least

at this time, about the transfer of a human embryo obtained by nuclear transfer techniques to a womb for purposes of gestation and birth.

Several conclusions emerge from this brief historical overview:

· Over the past twenty-five years, theologians have engaged in repeated discussions of the prospect of cloning humans that anticipate and illuminate much current religious discussion of this topic.
· Theological and ecclesiastical positions on cloning humans are pluralistic in their premises, their modes of argument, and even their conclusions. In short, they exhibit the pluralism characteristic of American religiosity.
· The religious discussion of cloning humans has connected it closely with ongoing debates about technologically assisted reproduction and genetic interventions.
· Despite changes in scientific research and technical capability, the *values* that underlie religious concerns about cloning humans have endured and continue to inform public debate.

Responsible Human Dominion Over Nature

Warnings Not To Play God. As often happens when a powerful new scientific tool is developed, the announcement that mammalian somatic cell nuclear transfer cloning was possible generated strong warnings against "playing God." This slogan is usually invoked as a moral stop sign to some scientific research or medical practice on the basis of one or more of the following distinctions between human beings and God:

· Human beings should not probe the fundamental secrets or mysteries of life, which belong to God.
· Human beings lack the authority to make certain decisions about

the beginning or ending of life. Such decisions are reserved to divine sovereignty.

· Human beings are fallible and also tend to evaluate actions according to their narrow, partial, and frequently self-interested perspectives.

· Human beings do not have the knowledge, especially knowledge of outcomes of actions, attributed to divine omniscience.

· Human beings do not have the power to control the outcomes of actions or processes that is a mark of divine omnipotence.

Even within religious communities, however, the warning against "playing God" may not be considered a sufficient argument against human cloning. Allen Verhey contends that this warning is simply too indiscriminate to provide ethical guidance. Furthermore, it overlooks moral invitations to play God, particularly in the realm of genetics (Verhey, 1995). While agreeing with Ramsey that human beings are not called to "play God," Protestant Ted Peters argues that this does not by itself define what is necessary for us to be human. Hence, we are responsible for using our creativity and freedom (features of the image of God) to forge a destiny more consonant with human dignity. In "playing human," Peters contends, there is no theological reason to leave human nature unchanged, and no theological principles that the cloning of humans necessarily violates (Peters, 1997).

Human Dignity

Appeals to human dignity are prominent in Roman Catholic analyses and assessments of the prospects of human cloning, which base "human dignity" on the creation story and on the Christian account of God's redemption of human beings. The Catholic moral tradition views the cloning of a human being as "a violation of human dignity" (Haas, letter from the Pope John Center, 1997).

Religious thinkers generally do not question whether a person created through cloning is a human being created in God's image. They extend to persons created through cloning the same moral protections that already apply to other persons created in the image of God. For instance, Rabbi Elliot Dorff argues that "[n]o clone may . . . legitimately be denied any of the rights and protections extended to any other child" (Dorff, 1997, p. 5). However, many fear that the human dignity of persons created through cloning will be violated by the denial of such rights and protections, for instance, through enslavement to others and other forms of "man's mastery over man" (Tendler, 1997).

Human cloning would violate human dignity, according to some religious opponents, because it would "jeopardize the personal and unique identity of the clone (or clones) as well as the person whose genome was thus duplicated" (Haas, 1997). This problem does not arise in the case of identical twins, because neither is the "source or maker of the other" (Haas, 1997). Religious concerns about identity and individuality focus mainly on how persons created through cloning will inevitably or possibly be treated, rather than whether such persons are actually unique creatures in God's image. Rejecting genetic determinism, religious thinkers hold that cloning humans would "produce independent human beings with histories and influences all their own and with their own free will" (Dorff, 1997, p. 6). The person created through cloning will be "a new person, an integrated body and mind, with unique experiences." However, it will doubtless be harder for such persons "to establish their own identity and for their creators to acknowledge and respect it" (Dorff, 1997, p. 6). Even for absolute opponents, the process of cloning humans only *violates* human dignity; it does not *diminish* human dignity: "In the cloning of humans there is an affront to human dignity. . . . Yet, in no way is the human dignity of that person [the one who results from cloning] diminished" (Haas, 1997, p. 3).

Sanctity of life is one norm associated with human dignity. For instance, the prohibition of the shedding of human blood is connected with God's creation of humans in his own image (Genesis 9:6). Opponents often view the cloning of a human as a breach, or at least as a potential breach, of the sanctity of life. In rejecting human cloning, Joseph Cardinal Ratizinger of the Vatican insisted that "the sanctity of [human] life is untouchable" (quoted in Haas, 1997, p. 2). Even those who offer limited support for human cloning, in part on the grounds that it could be used in support of life, argue that it is necessary to set conditions and limits in order to prevent harm to persons who are created through cloning. Not only do they rule out such egregious violations of the sanctity of life as sacrificing persons created through cloning in order to obtain their organs for transplantation, they also worry about what will be done with the "bad results," that is, the "mistakes" that will be inevitable at least in the short term (Dorff, 1997, pp. 3–4). In addition, most recognize that the risks to persons created through cloning are now so unknown that we should virtually rule out human cloning for the present, because those who create children in this manner could not be sure that they are "doing no evil" (Tendler, 1997).

Objectification also represents a fundamental breach of human dignity. To treat persons who are the sources of genetic material for cloning or persons who are created through cloning as mere objects, means, or instruments violates the religious principle of human dignity as well as the secular principle of respect for persons. Cloning humans would necessarily involve objectification, some religious thinkers argue, because it would treat the child as "an object of manipulation" by potentially eliminating the marital act and by attempting "to design and control the very identity of the child" (Haas, 1997). Cloning humans is wrong, in short, because "it subjects human individuals at their most vulnerable, at their very coming-into-being, to the arbitrary whim, power and manipulation of others" (Haas, 1997). For other religious thinkers who accept human cloning under some circumstances, it is necessary to reduce the effects of objectification, for example, by a commitment to accept and care for the "mistakes" made in cloning (Dorff, 1997).

Objectification can become commodification when commercial and economic forces determine whether and how a person is treated as an object. Religious opponents of human cloning stress that objectification through commodification is a major risk and worry that "economic incentives will control when humans will be cloned" (Cahill, 1997, p. 3). Commodification would deny "the sacred character of human life depicted in the Jewish tradition, transforming it instead to fungible commodities on the human marketplace to be judged by a given person's worth to others" (Dorff, 1997, p. 2).

Procreation and Families

Procreation and Reproduction. In the initial phase of theological debate about cloning humans, Paul Ramsey argued that the covenant of marriage includes the goods of sexual love and procreation, which are divinely ordained and intrinsically related: Human beings have no authority to sever what God had joined together. On this basis, Ramsey, a Protestant, joined with several Roman Catholic moral theologians, such as Bernard Häring and Richard Mc-Cormick, in objecting to the cloning of humans as part of the panoply of reproductive technologies. They claimed that such technologies separate the unitive and procreative ends of human sexuality and transform "procreation," which at most puts humans in a role of co-creator, into "reproduction." The Vatican's 1987 *Instruction on Respect for Human Life (Donum Vitae)* rejected human cloning either as a scientific outcome or technical proposal: "Attempts or hypotheses for obtaining a human being without any connection with sexuality through 'twin fission,' cloning, or parthenogenesis are to be considered contrary to the moral law, since they are in opposition to the dignity both of human procreation and of the conjugal union" (Congregation for the Doctrine of the Faith, 1987).

A similar critique distinguishes "begetting" (procreating) from "making" (reproducing). According to the Nicene Creed of early Christianity, Jesus, as the authentic image of God and the normative exemplar of personhood, is "begotten, not made" of God. The theological interpretation of "begetting" emphasizes likeness, identity, equality; begetting expresses the parent's very being. By contrast, "making" refers to unlikeness, alienation, and subordination; it expresses the parent's will as a project.

However, many religious thinkers do not accept the sharp separation between begetting and making, because it could rule out various reproductive technologies that they find acceptable, just as many do not accept the absolute connection between unitive and procreative meanings of sexual acts, in part because it would rule out artificial contraception, which they find acceptable. They may, nevertheless, still reject the cloning of humans to create children because they perceive it to be radically different from all other methods of technologically-assisted reproduction. Thus, they may stress the radically new features of human cloning, perhaps even viewing it as a "genuine revolution" in reproduction.

Concerns About the Family. Religious traditions usually approach the cloning of humans to create children from the standpoint of familial relationships and responsibilities rather than from the standpoint of personal rights and individual autonomy. Hence, a primary moral criterion is the impact of cloning humans on the integrity of the family, a concern that includes but also goes beyond the inseparable goods of marriage and the primacy of begetting over making.

Lisa Cahill, a Roman Catholic moral theologian, argues that "the child who is truly the child of a single parent is a genuine revolution in human history, and his or her advent should be viewed with immense caution." She further contends that cloning violates "the essential reality of human family and . . . the nature of the socially related individual within it. We all take part of our identity, both material or biological and social, from combined ancestral kinship networks. The existing practice of 'donating' gametes when the donors have no intention to parent the resulting child is already an affront to this order of things. But, in such cases, as in cases of adoption where the rearing of a child within its original combined-family network is impossible or undesirable, the child can still in fact claim the dual-lineage origin that characterizes every other human being. Whether socially recognized or not, this kind of an-cestry is an important part of the human sense of self (as witnessed by searches for 'biological' parents and families), as well as a foun-dation of important human relationships." Cloning humans to cre-ate children, Cahill concludes, would constitute an "unprecedented rupture in those biological dimensions of embodied humanity which have been most important for social cooperation" (Cahill, testimony, 1997). At the extreme, cloning humans would not only free human reproduction from marital and male-female relation-ships, but would "allow for the emancipation of human reproduc-tion from *any* relationship" (Mohler, 1997).

Concerns about lineage and intergenerational relations in other religious traditions also set limits on or challenge the cloning of hu-mans to create children. For example, Islamic scholar Abdulaziz Sachedina suggests that Islam could accept some therapeutic uses of human cloning "as long as the lineage of the child remains reli-giously unblemished" (Sachedina, 1997, pp. 6–7). And some Jew-ish thinkers worry that cloning humans may diminish the ethic of responsibility because of changed roles (father, mother, child) and relationships (spousal, parental, filial).

Assessments of Acts and Public Policies

Religious perspectives on public policies regarding human cloning vary for several reasons. One critical factor is whether the tradition views every possible act of cloning humans as intrinsically evil (as, for example, Roman Catholicism does) or whether it recognizes that cloning humans could conceivably be justified in some circumstances, however few they may be (as, for example, many in the Jewish tradition do). The Roman Catholic tradition argues that the very *use* of cloning techniques to create human beings is contrary to human dignity: "One may not use, even for a single instance, a means for achieving a good purpose which intrinsically is morally flawed" (Haas, 1997, p. 4). And, for that tradition, creating a child through human cloning is intrinsically morally flawed. Some thinkers in other traditions also hold that such an action is always morally wrong, whatever good might come from it (see Meilaender, 1997).

By contrast, some other religious thinkers believe that cloning a human to create a child could be religiously and morally acceptable under certain conditions. They may view the technology as "morally neutral" (Dorff, 1997) and then consider which uses are morally justified; or they may oppose human cloning from matured (differentiated) cells except in the most exceptional circumstances and then identify those exceptional circumstances.

Two hypothetical scenarios are quite common. The first one involves cloning a sterile person to create a child. Rabbi Tendler poses the case of "a young man who is sterile, whose family was wiped out in the Holocaust, and [who] is the last of a genetic line." Rabbi Tendler says "I would certainly clone him" (Tendler, 1997, transcript, p. 35). The debate about this type of case hinges in part on different views of infertility. The Jewish tradition often views infertility as an "illness" and thus brings it under the responsibility to heal. According to others, for example, some in the Protestant tradition, the problem of infertility is not serious enough to warrant research into or actual human cloning (see Duff, 1997, p. 5).

A second case involves cloning a person who has a serious and perhaps fatal disease and needs a compatible source of biological material, such as bone marrow. Rabbi Dorff, for instance, holds that it would be "legitimate from a moral and a Jewish point of view" to clone a person with leukemia with the intent of transplanting bone marrow from the created child as long as the "parents" intend to raise the child as they would raise any other child (Dorff, 1997, pp. 4–5; see also Tendler, 1997). Some Protestants concur on this case, even when they reject the first type of case (see Duff, 1997, p. 4). Those who consider the second type of case justifiable rule out destruction or abandonment of the created child, as well as the imposition of serious risks of harm. Indeed, acceptance of either type of hypothetical case—as well as a third type of case involving the cloning of a dying child—presupposes that the procedure is safe for the child created by cloning. Other conditions include the protection of the created child's rights and the lack of acceptable alternatives to cloning persons in such cases.

Those who view cloning humans as intrinsically wrong may also respond sympathetically and compassionately to people's suffering when they are infertile or have a disease that brings death or disability. However, they usually hold that the good of overcoming this suffering does not justify cloning humans: Cloning "is entirely unsuitable for human procreation even for exceptional circumstances" (Haas, 1997, p. 4). Indeed, religious critics may view the exceptional circumstances featured in the cases as "temptations" to be resisted (see Meilaender, 1997, p. 5).

Some rough correlations hold between evaluations of particular cases and proposals for public policy. Religious thinkers who view the cloning of a human being as intrinsically wrong, i.e., wrong in and of itself, under any and all circumstances, tend to support a permanent ban on cloning humans through legislative and other means. Any use of cloning technology to create a human child abuses that technology, which is, however, acceptable in animal reproduction. By contrast, religious thinkers who hold that, in some conceivable circumstances, it could be morally justifiable to clone a person to create a child tend to support public policies that regulate the procedure, with varying restrictions, or that ban the procedure for the time being or until certain conditions are met. In assessing public policies, this second group is particularly concerned to prevent potential abuses of the technology in cloning humans rather than condemning all uses.

Most religious thinkers who recommend public policies on cloning humans propose either a ban or restrictive regulation. A few examples will suffice. On March 6, 1997, the Christian Life Commission of the Southern Baptist Convention issued a resolution entitled "Against Human Cloning," which supported President Clinton's decision to prohibit federal funding for human-cloning research and requested "that the Congress of the United States make human cloning unlawful." The resolution also called on "all nations of the world to make efforts to prevent the cloning of any human being."

The Vatican's 1987 *Instruction on Respect for Human Life (Donum Vitae)* argued for a legal prohibition of human cloning, as well as many other reproductive technologies. Official Roman Catholic statements since that time have condemned nontherapeutic research on human embryos and human cloning and have called on governments around the world to enact prohibitive legislation. Most recently, in the wake of the cloning of Dolly, a Vatican statement reiterated the basic teaching of *Donum Vitae:* "A person has the right to be born in a human way. It is to be strongly hoped that states . . . will immediately pass a law that bans the application of cloning of humans and that in the face of pressures, they have the force to make no concessions."

By contrast, Rabbi Elliot Dorff argues that "human cloning should be regulated, not banned." He holds that "the Jewish demand that we do our best to provide healing makes it important that we take advantage of the promise of cloning to aid us in finding cures for a variety of diseases and in overcoming infertility." However, "the dangers of cloning . . . require that it be supervised and restricted." More specifically, "cloning should be allowed only for medical research or therapy; the full and equal status of clones with other fetuses or human beings must be recognized, with the equivalent protections guarded; and careful policies must be devised to determine how cloning mistakes will be identified and handled" (Dorff, 1997). Although Dorff stresses legislation, particularly to regulate privately funded research, he recognizes that legislation will be only partially effective, and for that reason calls for increased attention to hospital ethics committees and institutional review boards, in part because of the self-regulation involved. Hence, although legislation is important "to ban the most egregious practices," most supervision "should come from self-regulation akin to what we already have in palce for experiments on human subjects" (Dorff, 1997, p. 15).

Conclusions

The wide variety of religious traditions and beliefs epitomizes the pluralism of American culture. Moreover, religious perspectives on cloning humans differ in fundamental premises, modes of reasoning, and conclusions. As a result, there is no single "religious" view on cloning humans, any more than for most moral issues in biomedicine. Nevertheless, discourse on many contested issues in biomedicine still proceeds across religious traditions, as well as secular traditions. Specifically with regard to cloning humans to create children, some religious thinkers believe that this technology could have some legitimate uses and thus could be justified under some circumstances if perfected; however, they may argue for regulation because of the danger of abuses or even for a ban, perhaps temporary, in light of concerns about safety. Other religious thinkers deny that this technology has any legitimate uses, contending that it always violates fundamental moral norms, such as human dignity. Such thinkers often argue for a legislative ban on all cloning of humans to create children. Finally, religious communities and thinkers draw on ancient and diverse traditions of moral reflection to address the cloning of humans, a subject they have debated off and on over the last thirty years. For some, fundamental religious beliefs and norms provide a clear negative answer: It is now and will continue to be wrong to clone a human. Others, however, hold that more reflection is needed, given new scientific and technological developments, to determine exactly how to interpret and evaluate the prospect of human cloning in light of fundamental religious convictions and norms.

References

Cahill, L. S., "Cloning: Religion-Based Perspectives," Testimony before the National Bioethics Advisory Commission, March 13, 1997.

Congregation for the Doctrine of the Faith, *Instruction on Respect for Human Life in Its Origin and on the Dignity of Procreation* (Rome, 1987).

Dorff, R. E. N., "Human Cloning: A Jewish Perspective," Testimony before the National Bioethics Commission, March 14, 1997.

Duff, N. J., "Theological Reflections on Human Cloning," Testimony presented to the National Bioethics Advisory Commission, March 13, 1997.

Fletcher, J., *Humanhood: Essays in Biomedical Ethics* (Buffalo, NY: Prometheus Books, 1979).

Fletcher, J., *The Ethics of Genetic Control* (Garden City, NY: Anchor Press, 1974).

Fletcher, J., "New beginnings in human life: A theologian's response," *The New Genetics and the Future of Man,* M. Hamilton (ed.) (Grand Rapids, MI: Wm. B. Eerdmans Publishing Company, 1972, 78–79).

Fletcher, J., "Ethical aspects of genetics controls," *New England Journal of Medicine* 285(14):776–783, 1971.

Haas, J. M., letter from the Pope John Center, submitted to the National Bioethics Advisory Commission, March 31, 1997.

Lynn, B., *Genetic Manipulation* (New York: Office for Church in Society, United Church of Christ, 1977).

McCormick, R. A., "Blastomere separation: Some concerns," *Hastings Center Report* 24(2):14–16, 1994.

McCormick, R. A., "Should we clone humans?," *The Christian Century* 17–24:1148–1149, November 1993.

Meilaender, G. C., Testimony before the National Bioethics Advisory Commission, March 13, 1997.

Mohler, R. A., "The Brave New World of Cloning: A Christian Worldview Perspective," (unpublished manuscript, March 1997).

Peters, T., *Playing God? Genetic Discrimination and Human Freedom* (New York: Routledge, 1997).

Ramsey, P., "Moral and Religious Implications of Genetic Control," *Genetics and the Future of Man,* John D. Roslansky (ed.) (New York: Appleton-Century-Croffs, 1966).

Ramsey, P., *Fabricated Man: The Ethics of Genetic Control* (New Haven: Yale University Press, 1970).

Rorvik, D., *In His Image: The Cloning of a Man* (Philadelphia: J.B. Lippincott Company, 1978).

Sachedina, A., "Islamic Perspectives on Cloning," Testimony before the National Bioethics Advisory Commission, March 14, 1997.

Tendler, R. M., Testimony before the National Bioethics Advisory Commission, March 14, 1997.

Verhey, A., "Playing God and invoking a perspective," *Journal of Medicine and Philosophy* 20:347–364, 1995.

This is an abridged version of chapter 3 of NBAC report on human cloning (June 1997). Much of the material in the original version is derived from a commissioned paper prepared for the National Bioethics Advisory Commission by Courtney S. Campbell, Department of Philosophy, Oregon State University, titled "Religious Perspectives on Human Cloning."

To Clone or Not to Clone

Jean Bethke Elshtain

Cloning is upon us. The techno-enthusiasts in our midst celebrate the collapse of yet another barrier to human mastery and control. But for most of us, this is an extraordinarily unsettling development. Talk to the man and woman in the street and you hear murmurs and rumblings and much dark musing about portents of the end-times and "now we've gone too far." The airwaves and the street win this one hands down, a welcome contrast to the celebratory glitz of *USA Today* trumpeting "Hello Dolly!"—Dolly being the name of the fetching ewe that faced the reader straight-on in a front page color photo announcing her cloned arrival. The sub-head read, "Sheep cloning prompts ethical debate." The sheep looked perfectly normal, of course, and not terribly exercised about her historic significance. That she was really the child of no one—no one's little lamb—will probably not haunt her nights and bedevil her days. But we—we humans—should be haunted, by Dolly and all the Dollies to come and by the prospect that others are to appear on this earth as the progeny of our omnipotent striving, our yearning to create without pausing to reflect on what we are destroying.

When I pondered cloning initially, a Chicago Bulls game was on television. The Bulls were clobbering the Spurs. Michael Jordan had just performed a typically superhuman feat, an assist that suggested he has eyes in the back of his head and two sets of arms. To one buoyant citizen—a rare optimist among the worriers—who called a local program to register his two cents worth on cloning, the prospect of "more Michael Jordans" made the whole "cloning thing" worthwhile. "Can you imagine a whole basketball team of Michael Jordans?" he queried giddily. Unfortunately, I could. It seemed to me then and seems to me now a nightmare. If there were basketball teams fielding Jordans against Jordans, we wouldn't be able to recognize the one, the only, Michael Jordan. It's rather like suggesting that forty Mozarts are better than one. But there would be no Mozart were there forty Mozarts. We know the singularity of the one; the extraordinary genius—a Jordan, a Mozart—because they stand apart from and above the rest. Absent that irreducible singularity, their gifts and glorious, soaring accomplishments would come to mean nothing as they would have become the norm, just commonplace. Another dunk; another concerto. In fact, lots of callers made this point, or one similar to it, reacting to the Michael Jordon Clontopia scenario.

A research librarian at a small college in Indiana, who had driven me to her campus for the purpose of delivering a lecture, offered a spontaneous, sustained, and troubled critique of cloning that rivals the best dystopian fictions. Her cloning nightmare was a veritable army of Hitlers, ruthless and remorseless bigots and killers who kept reproducing themselves and were one day able to finish what the historic Hitler failed to accomplish. It occurred to me that an equal number of Mother Theresas would probably not be a viable deterrent, not if the Hitler clones were behaving like, well, Hitlers.

But I had my own nightmare scenario to offer. Imagine, I suggested to my librarian driver, a society that clones human beings to serve as spare parts. Because the cloned entities are not fully human, our moral queasiness could be disarmed and we could "harvest" organs to our heart's content—and organs from human beings of every age, race, phenotype at that. Harvesting organs from anencephalic newborns would, in that new world, be the equivalent of the Model T—an early and, it turns out, very rudimentary prototype of glorious, gleaming things to come.

Far-fetched? No longer. Besides, often the far-fetched gets us nearer the truth of the matter than all the cautious, persnickety pieces that fail to come anywhere close to the pity and terror this topic evokes. Consider Stanislaw Lem's *The Star Diaries,* in which his protagonist, Ijon Tichy, described as a "hapless Candide of the Cosmos," ventures into space encountering one weird situation after another. Lem's "Thirteenth Voyage" takes him to a planet, Panta, where he runs afoul of local custom and is accused of the worst of crimes, "the crime of personal differentiation." The evidence against him is incriminating. Nonetheless, Tichy is given an opportunity to conform. A planet spokesman offers a peroration to Tichy concerning the benefits of his planet, on which there are no separate entities—"only the collective."

For the denizens of Panta have come to understand that the source of all "the cares, sufferings and misfortunes to which beings, gathered together in societies, are prone" lies in the individual, "in his private identity." The individual, by contrast to the collective, is "characterized by uncertainty, indecision, inconsistency of action, and above all—by impermanence." Having "completely eliminated individuality," on planet Panta they have achieved "the highest degree of social interchangeability." It works rather the way the Marxist utopia was to function. Everyone at any moment can be anything else. Functions or roles are interchangeable. On Panta you occupy a role for twenty-four hours only: one day a gardener, the next an engineer, then a mason, now a judge.

The same principle holds with families. "Each is composed of relatives—there's a father, mother, children. Only the functions remain constant; the ones who perform them are changed every day." All feelings and emotions are entirely abstract. One never needs to grieve or to mourn as everyone is infinitely replaceable. "Affection, respect, love where at one time gnawed by constant anxiety, by the fear of losing the person held dear. This dread we have conquered. For in point of fact whatever upheavals, diseases or calamities may be visited upon us, we shall always have a father, a mother, a spouse, and children." As well, there is no "I." And there can be no death "where there are no individuals. We do not die." Tichy can't quite get with the program. Brought before a court, he is "found guilty and condemned to life identification." He blasts off and sets his course for Earth.

Were Lem writing an addendum for his brilliant tale, he might show Tichy landing, believing he is at last on terra firma in both the literal and metaphorical sense, only to discover that the greeting party at the rocket-port is a bit strange: There are forty very tall basketball players all in identical uniforms wearing No. 23 jerseys, on one side and, on the other, forty men in powdered wigs, suited up in breeches and satin frock coats and playing identical pieces on identical harpsichords. Wrong planet? No more.

Sure, it's amusing, up to a point. But it was anything but amusing to overhear the speculation that cloning might be made available to parents about to lose a child to leukemia or, having lost a child to an accident, in order that they might reproduce and replace that lost child. This image borders on an obscenity. Perhaps we need a new word to describe what it represents, to capture fully what order of things the cloning of children in order to forestall human loss and grief violates. We say to little Tommy, in effect: "Sorry to lose you. But Tommy 2 is waiting in the wings." And what of Tommy 2? What happens when he learns he is the pinch hitter? "There was an earlier Tommy, much loved, so Mommy and Daddy had a copy made." But it isn't really Mommy and Daddy— it's the two people who placed the order for him and paid a huge sum. He's their little product; little fabricated Tommy 2, a techno-orphan. And Tommy 1 lies in the grave unmourned; undifferentiated in death; unremembered because he had been copied and his individuality wrenchingly obliterated.

The usual nostrums are of no use here. I have in mind the standard cliché that, once again, our "ethical thinking" hasn't caught up with technological "advance." This is a flawed way to reflect on cloning and so much else. The problem is not that we must somehow catch our ethics up to our technology. The problem is that technology is rapidly gutting our ethics. And it is *our* ethics. Ethical reflection belongs to all of us—all those agitated radio callers—and it is the fears and apprehension of ordinary citizens that should be paid close and respectful attention. The ethicists are cut from the same cloth as everybody else. They breathe the same cultural air. They, too, are children of the West, of Judaism, Catholicism, the Renaissance, the Reformation, the Enlightenment. In the matter of cloning, we cannot wait for the experts. The queasiness the vast majority of Americans feel at this "remarkable achievement" is appropriate and should be aired and explored fully.

Perhaps something remarkable will finally happen. We will put the genie back into the bottle for a change. We will say, "No, stop, we will not go down this road." This doesn't make us antiscience or antiprogress or stodgy sticks-in-the-mud. It makes us skeptical, alert, and, yes, frightened citizens asking the question: Whatever will become of the ancient prayer, "That I may see my children's children and peace upon Israel," in a world of cloned entities, peopled by the children of No Body, copies of our selves? These poor children of our fantasies and our drive to perfect and our arrogant search for dominion: What are we to say to them? Forgive us, for we knew not what we were doing? That tastes bitter on the tongue. We knew what we were doing and we did it anyway. Of whom will we ask forgiveness? Who will be there to listen? Who to absolve?

Are these the musings of an alarmist, a technophobe, a Luddite? Consider that there are now cloned calves in Wisconsin and cloned rodents in various laboratories worldwide. Cloned company is bursting out all over: thus far none of it human. The clone enthusiasts will surely find a way, however. Dolly's creator or producer or manufacturer—hard to know what to call him—thinks human cloning is a bad idea. But others are not nearly so reticent. Consider, then, some further developments on the cloning and related

186 JEAN BETHKE ELSHTAIN

fronts that promise, or threaten, to alter our relation to our bodies, our selves.

A big story of the moment—and a huge step toward human cloning—lies in the fertile field of infertility science: the world of human reproductive technology. Many procedures once considered radical are by now routine. These include in vitro fertilization, embryo flushing, surrogate embryo transfer, and sex preselection, among others. Now comes Dr. Mark Sauer, described by the *New York Times* as "an infertility expert at Columbia Presbyterian Medical Center in New York" who "dreams of offering his patients a type of cloning some day." It would work like this. You take a two- or three-day-old human embryo and use its cells—there are only about eight at this stage—to grow identical embryos where once there was only one. The next step is to implant "some" of these embryos in a woman's uterus immediately and freeze the extras. And what are the plans for the clonettes in cold storage? Well, initial attempts at impregnation may fail. So you have some spare embryos for a second, third, or fourth try. Suppose the woman successfully carries the initial implants to term. She may want more babies—identical babies—and the embryos are there for future use. The upshot, of course, is that a woman could wind up with "identical twins, triplets, or even quadruplets, possibly born years apart."

And why would anyone want this, considering the potentially shattering questions it presents to the identity and integrity of the children involved? Dr. Sauer has an answer. Otherwise there "might be no babies at all." To be sure, the premise of this procedure isn't as obviously morally repugnant as the scenario noted above, the speculation that cloning might be made available to parents about to lose a child to leukemia or, having lost a child to an accident, in order that they might reproduce and replace that child, as I noted already.

Rather, the debate about this latest embryo cloning scenario, by contrast, rages around whether or not this is, in fact, cloning at all or whether it is a version of cloning that is more or less question-

able than the standard or classic form: the Dolly scenario. Dr. Sauer
and other enthusiasts say that because cloning is a "politically dirty
word"—there is, apparently, no real ethical issue here—they hope
that their proposed method of crypto-cloning may slip under the
radar screen. Besides, he avers, it's much better for the women involved: You don't have to give them lots of drugs to "force their
ovaries to pump out multiple eggs so that they could fertilize them
and create as many embryos as possible."

Again, why are so many women putting themselves through
this? And why has this been surrounded by the halo of "rights"?
You can be sure, once word gets around, that the more "attractive"
idea (in the words of another infertility specialist) of replicating
embryos will generate political demands. A group will spring up
proclaiming "embryo duplication rights" just as an outfit emerged
instantly after Dolly was announced arguing that to clone oneself
was a fundamental right. Several of the infertility specialists cited
in the *Times* piece, all male doctors, interestingly enough, spoke of
the pleading of women, of "the misery my patients are living
through." But surely a good bit of that misery comes from having
expectations lifted out of all proportion in relation to chances of
success (with procedures like in vitro), only to find, time and time
again, that the miracle of modern medicine has turned into an invasive, expensive, mind-bending, heart-rending dud. A doleful de-
nouement to high-tech generated expectations and the playing out
of "reproductive freedom."

Whatever happened to accepting embodied limits with better
grace? There are many ways to enact what the late Erik Erikson
called "generative" projects and lives. Biological parenthood is one
but not the only one. Many of the women we call great from our
own history—I think here of one of my own heroes—Jane Addams
of Hull-House—were not mothers although they did an extraordinary amount of mothering. Either through necessity or choice,
she and many others offered their lives in service to civic or religious projects that located them in a world of relationships over the

years with children not their own that involved loving concern, care, friendship, nurture, protection, discipline, pride, disappointment: all the complex virtues, habits, and emotions called forth by biological parenting.

And there is adoption, notwithstanding the frustrations many encounter and the fear instilled by such outrageous violations of decency as the holding in the "Baby Richard" and other recent cases in which children were wrenched from the only family they had ever known in order to be returned to a bio-parent claimant who had discovered belatedly the overwhelming need to be a father or mother. How odd that biology now trumps nearly all other claims and desires. In several texts I've encountered recently, adoption is surrounded with a faintly sinister odor and treated as an activity not all that different from baby selling. Somehow all these developments—the insistent urge to reproduce through any means necessary and the emergence of a multimillion-dollars-a-year specialty devoted to precisely that task; the diminution of the integrity of adoption in favor of often dubious claims from bio-parents; the possibility, now, of cloning embryos in order to guarantee more or less identical offspring to a desperate couple—are linked.

What common threads tie these disparate activities together? How does one account for the fact that the resurgence of feminism over the past thirty years and enhanced pressures on women, many of them placed on women by themselves, to reproduce biologically have emerged in tandem? Why are these developments surrounded by such a desperate aura and a sense of misery and failure—including the failure of many marriages that cannot survive the tumult of infertility high-tech medicine's intrusion into a couple's intimate lives? Let's try out one possible explanation. Here at the end of the twentieth century we all care mightily about identity: who we are. Sometimes this takes the form of identity politics in which one's own identity gets submerged into that of a group, likely a group defined in biological or quasi-biological terms on grounds of sex, race, or ethnicity. That's problematic enough as a basis for politics, to say the least. But we've further compounded the biological urgencies, upping the ante to bear one's "own" child as a measure of the success or failure of the self.

Mind you, I do not want to downplay how heartbreaking it is for many couples who want to have a baby and cannot. But, again, there are many ways to parent and many babies desperate for loving families. Rather than to expand our sense of gracious acceptance of those who may not be our direct biological offspring, which means accepting our own limits but coming to see that these open up other possibilities, we rail against cruel fate and reckon ourselves nigh-worthless persons if we fail biologically. Perhaps with so much up for grabs, in light of the incessant drumbeat to be all we can be, to achieve, to produce, to succeed, to define our own projects, to be the sole creators of our own destinies, we have fallen back on the bedrock of biology. When all that is solid is melting into air, maybe biology seems the last redoubt of solidity, of identity. But, of course, this is chimerical. In demanding of our bodies what they sometimes cannot give, our world grows smaller, our focus more singular if not obsessive, and identity itself is called into question: our own and that of our future, identical offspring.

This article appeared, in slightly different form, as "The Hard Questions: Our Bodies, Our Clones," in the *New Republic,* August 4, 1997, pages 25 and following.

Human Cloning and the Public Realm: A Defense of Intuitions of the Good

David Tracy

The Dilemma

In 1947 a surprising discovery was made: The United Nations, in its desire to enlist cross-cultural agreements on basic human rights, appointed a committee of ethical, political, and religious thinkers to determine what rights did cross-cultures have and for what ethical, political, metaphysical, or religious reasons. A consensus on certain basic human, civic, and political rights emerged. But there proved to be no way for any philosopher to win agreement on anything like a common ethical, political, metaphysical, or religious answer to the question of *why* the most basic human rights were just that: basic human rights. Fortunately, the U.N. Declaration on Human Rights (1948) was passed anyway. Liberals agreed—but for their own ethical-political reasons; the same with Marxists, conservatives, and radicals. Jews and Christians, Muslims, Buddhists, Hindus, Taoists, Confucianists, and peoples of several indigenous religions found it possible to agree with the practical list of basic human rights—but each for their own ethical, metaphysical, or religious reasons. Desperate as the situations on human rights remain in many places of our globe (e.g., China, Sudan, North Korea, etc.) we are all far better off for that earlier U.N. Declaration on Human Rights even if a more far-ranging consensus on the fundamental ethical reasons for those rights is still lacking—and is likely to remain lacking for the foreseeable future. At least the torturers may sleep a little less easily; at least some of the torture, false imprisonments, assaults on ethnic, racial, gender, civic, religious, and individual rights have been slowed down, although clearly not halted. At least there is a consensus document that all relevant parties agreed upon and can be held accountable to.

Many philosophers and ethicists at the time (including several who participated in the writing of the document, such as Jacques Maritain) lamented their failure to reach any common agreement on the reasons why certain human rights were indeed basic. Almost fifty years later, there is still no agreement on *why* certain rights are the most basic human rights. Does anyone seriously think that any agreement could possibly be achieved today on the much more complex ethical issues at stake in the present debate on human cloning? After all, that debate demands reflection not just on human rights but on every meaning of what constitutes an authentic human being. Is it possible in our situation, so much more pluralistic, complex, and global than that of 1948, to hope for any consensus at all this time, even on the practical step needed: a ban? a moratorium? research full speed ahead? The only serious hope is to increase the range of conversation-partners to the discussion: first, to educate ourselves as best we nonscientists can in the complex scientific procedures, techniques, and facts involved (here the media—at least the major print media—have performed admirably to inform us in lay terms); second, to endorse President Clinton's charge to his commission that all informed parties should join in the discussion.

Any philosophical, ethical, or religious individual or tradition that can help focus the discussion on human cloning by rendering available the ethical resources of their traditions should be welcome to this crucial and unavoidable discussion. None of us, to be sure, will be much aided by those familiar factions, secular and religious alike, who continue to function like "certainty-factories" with their quick, ready-made pronouncements based on clear and distinct ideas and very few of them.

If ever there was an issue that demanded both a sense for intellectual complexity and ethical ambiguity, human cloning is that issue. For somehow we must find a way together to go back to where the debates on human rights left off, in the hope of uncovering our most basic intuitions of what we ultimately believe to be human. We must be willing to force ourselves to try to articulate our basic moral feelings, emotions, intuitions on the human. What Albert Einstein famously said about the atomic age—"Everything has changed except our thinking"—is even more true on the issues raised by human cloning. How can we think well and responsibly without hearing all the voices and traditions that deserve to be heard?

Any reading of the growing literature on the possibility of human cloning (President Clinton's clear and surprisingly strong statement, his commission's first report, the many institutional or individual essays of the last year) shows that there is clearly no real consensus. Possibly there never will be. But there is enough consensus, I believe, on certain shared deeply troubling questions and moral intuitions raised.

1. Do we really want the emerging biotechnology, so overwhelming in both its promise and threat for the future, to proceed without serious ethical reflection? Clearly no. The danger is not technology nor biotechnology; the danger is an emerging market-driven biotechnocracy that is as dangerous (because it is as unthinking) as any other totality system of the past.

2. Do we acknowledge that in the emerging global monoculture every significant human cultural difference and otherness may be destroyed as the quiet regimes of economic and political power (indeed more the economic than the political) find ever more effective market strategies to enforce the rule of what Michael Foucault nicely called the reign of "more of the same"? Human cloning certainly sounds like the ultimate contribution to an undesirable monoculture.

3. Are we not justly alarmed at the disturbing lack of agreement on some basic ethical understanding of what constitutes a human person as human? If the global market alone dictates the future on human cloning, the answers to this question will eventually drown out every other voice. As many of our most serious social critics have argued, we are already too inflicted by "possessive individualism." Would human beings desire to become what we have already made of the nonhuman world: manufactured and marketable commodities?

4. Does not the present debate have the danger of excluding sufficient reflection on the increasing gap between the rich and powerful countries and the poor and relatively powerless ones? If we already have a situation where the poor of the world sometimes have few options except to sell body organs for the health of the rich, where would the realities of human cloning take us? Would it become (as it easily could if only the market decides) a new luxury item for the rich, the beautiful, the talented, the famous?

5. On the other hand, opponents of human cloning (as I am) cannot afford to ignore the benefits that such cloning *might* provide for all humankind, for example, in helping to control or eliminate some genetic diseases, or as a possibility needed in some extreme situations such as the only viable alternative to survival if some literally uncontrollable virus were let loose on the globe? Or yet more speculatively, the late Carl Sagan is persuasive that it is more probable than not that there is intelligent life elsewhere in the universe. What if, in the future, the only way to contact those "others," given the limitations of our present bodies, was to devise new human bodies for the presently inconceivable journey to other galaxies? However, as several scientific commentators have observed, science may be able, with research already in progress, to solve some, if not all, of these dilemmas without crossing the moral

chasm of human cloning. Is that the case? What should we do if it were not the case?

President Clinton's commission is clearly wise to call for at least a moratorium on human cloning until the scientific facts are clear and the ethical consequences are widely discussed. Now the pressing issue is how to discuss these matters in a public way.

The Public Realm: Arguments and Intuitions of Human Goods

Consider the contemporary discussion of the nature of publicness itself. In a pluralist culture, it is important to know what will and will not count as public—that is, available to all intelligent, reasonable, and responsible members of that culture despite their otherwise crucial differences in belief and practice. A public realm assumes that there is the possibility of discussion (argument, conversation) among all participants. The only hope for such discussion in a radically pluralist culture is one based on reason. But today to state that reason is the solution is to restate the problem of publicness, not to resolve it.

At least this much seems clear on what seems to constitute a public realm. To produce public discourse is to provide reasons for one's assertions. To provide reasons is to render one's claims shareable and public. To provide reasons is to be willing to engage in argument. Argument is the most obvious form of public discourse. To engage in argument is minimally to make claims and to give the warrants and backings for those claims.

The move to explicit argument is the most obvious way to ensure publicness. If there is a public realm at all, this means at least that there is a space where argument is not merely allowed but demanded of all participants. This means, as well, that truth in the public realm will be fundamentally a matter of consensus—a consensus of the community of inquiry cognizant of and guided by the criteria and evidence of whatever the particular subject matter under discussion demands. A community of inquiry must be democratic, even radically egalitarian, in the most fundamental sense: the sense that no one can be accorded privileged status in an argument; all are in principle equal; all are bound to produce and yield to evidence, warrants, backing. Any emerging consensus must be a consensus responsible to the best argument on both the scientific and ethical questions at stake.

The first responsibility of the public realm, therefore, is the responsibility to give reasons, to provide arguments—to be public. Argument has traditionally been, and must remain the primary candidate for publicness. And yet there is a second candidate as well: one related to, yet distinct from argument itself. That candidate is an inquiry into various intuitions of the good, including those expressed in art and religion.

Religions, for example, characteristically provide responses to questions at the limits of human argument and even human experience. These religious questions—these limit-questions, if you will—remain relatively stable across the wide and often conflicting responses of the religions. Since we do not really receive answers to questions we have never asked, it is important to find disciplined ways to formulate the peculiar kinds of questions, experiences, and intuitions to which religions typically appeal. Indeed, such questions abound for any thoughtful person: What, if anything, is the meaning of the whole? What, if any, is the significance of such positive experiences as a fundamental trust empowering the fact that we continue to go on at all, or such negative experiences as a fundamental anxiety in the face of no specific object (No-thing) as distinct from fear in the face of some specific object? What is our primordial intuitive response to finitude, to contingency, to mortality, to radical oppression or alienation, to joy, love, wonder, and those strange experiences mystics describe as a consolation without a cause? What do we ultimately feel, sense, intuit, think that a human being is as human?

What is the meaning of the fact that our best reflective enter-

prises seem to disclose limits at the edge of their argumentative inquiry which can seem to suggest some other dimension, perhaps
even some glimpse of the character of the whole: the realization,
for example, that some intelligible order must exist in order for scientific inquiry to function at all; the disturbing question of why
be moral at all at the very limit of all our moral convictions? We
may reasonably call all such genuinely religious questions "limit-
questions."

To choose the category "limit" to describe the kind of questions
that religions address is to recall, of course, Kant's definition of
limit as "that which can be thought but not known." Insofar as we
try to describe what can be thought but not known, we do not
need to insist that the discussion employ the Western religious
term of "God" or the Western philosophical category of the "Absolute." We can choose, as I did above, the more flexible and admittedly more vague category of "the whole" and thus find at least
some initial way to use the Western category of "limit" without pre-
cluding its use for the intuitions on the whole of non-Western religions as well. To be able to sense some intuition of the whole, even
when we cannot know the whole, suggests anew, as Emerson saw
so clearly, the call within any reasonable person to allow for distinct
modes of inquiry upon all human limit-experiences and limit-
questions.

If we are willing to risk an interpretation of the religions for a
discussion of basic intuitions on the good, moreover, we must also
acknowledge that the risk is inevitably great. For the religious phenomenon is a deeply ambiguous phenomenon in human thought
and history. It is likely to remain so. Religion is cognitively ambiguous as necessarily approached and expressed indirectly (e.g.,
through limit-language). That cognitive ambiguity often yields positive, if often indirect and symbolic, intuitive fruit for thought and
life—as does religion's most natural analogue, art. Yet that cognitive ambiguity can also yield such negative intellectual fruits as ir-

rationality, obscurantism, and mystification with their attendant intellectual and ethical damage.

Religions release not only great creative possibilities for the good in individuals, societies, and whole cultures; religions also release frightening, even demonic realities—as the history of religion in any culture shows. And yet this cognitive and ethical ambiguity of religion, with its disclosure of the true and the false, the good and the evil, even the "beyond good and evil" possibilities of the holy, should be sufficient evidence to warrant the belief that religions are crucial phenomena for all in the public realm to risk interpreting. In the discussion on human cloning we should interpret religions as fundamentally intuitions and visions of the good. Thus interpreted, the religions could teach much about some of our most basic intuitions on a possibly shared humanity, especially on the central question of what constitutes a human being as human. That question eventually becomes a question not solely of rights (as it must be) but of some visions of the human good.

The kind of cultural pluralism that already exists in the contemporary public realm is matched by a similar pluralism in contemporary religion and art, in their sometimes complementary, sometimes conflicting intuitions and visions of the human good. In principle, pluralism is an enriching, not an impoverishing reality. In fact, pluralism is often an unnerving reality. For unless we learn to converse better and argue more clearly with one another on how to provide better descriptions of and reflection upon our distinct visions of the human good, we are all in danger of allowing the promise of cultural and religious pluralism to slide into a kind of Will Rogers pluralism—one where you never met an opinion you didn't like. Any responsible pluralist has met unacceptable opinions and intuitions and, when pressed, should be able to state clearly just why this opinion is wrong. As Isaiah Berlin, one of the great defenders of pluralism in politics and culture, once observed, a responsible pluralist will always be able to tell the better from the good and the good from the bad and the downright awful. That is the kind of pluralism needed for a public discussion on intuitions and visions of the human good when facing the communal questions and unavoidable practical communal decisions demanded by the debate on the possibility of human cloning.

Religions as Intuitive Visions of the Good

Any contemporary discussion of intuitions, feelings, and visions of the good that will bear public use must be articulated with as much philosophical care as the subject matter allows. This is even more the case, as Hans Jonas argued, in a situation where technology has so great a role in forming and transforming our personal and societal intuitions, feelings, and visions of the good for human beings. Two great resources for discussing some of our most basic intuitions of the good are clearly art and religion, both read here as expressions, personal and communal, of intuitions and visions of the human good.

The difficulties for a fruitful discussion are, however, also quite clear: (1) Given the increasing power of the techno-economic realm (i.e., technological innovations driving and driven by the global market economy), even "reason" can become merely technical reason, that is, capable of careful formal arguments on efficient means and, at its substantive best, on rights and procedures. But how does reason, thus narrowed, discuss, as it once did, not merely means but ends—including intuition of ends as a human good? Even defenders of the pluralistic democratic liberal theory for society (as I am) can become alarmed that the discussion of "goods," not merely "rights," is relatively impoverished in modern liberal political theory. Some of the contemporary debate on cloning sometimes reads as if the hands are still the hands of John Stuart Mill but the unintended and subconscious voice is that of Dr. Mengele.

There must be better ways to visions of the human good than most liberal democratic theory presently allows. If reason is rendered merely technical, art is sure to become marginalized and religion privatized. Of course, in a pluralistic, democratic society, everyone is welcome to live with her or his vision of the good. But preferably they should live on what Adorno called a "reservation of the spirit." For the public debate too often excludes all public debates on intuitions and visions of the good (or "ends") and thereby the use of all the cultural and symbolic resources of art and religion, except of course as "private" visions of private individuals or communities. Indeed, even without the aid of either art or religion, all of us presumably have learned in the last ten years at least this much from the shocking revelations of the extent of child abuse and spousal abuse: We do in fact share a repulsion, a moral outrage at such conduct as unacceptable for anyone claiming to be a human being respectful in the most fundamental human sense of other human beings.

Fortunately, the feelings and intuitions of the good in art now have distinguished public defenders in such thoughtful philosophers as Iris Murdoch, Martha Nussbaum, Charles Taylor, even, at times Richard Rorty. But the resources of religions on visions of the good can seem a far more dangerous choice for entry into the discussion. Of course, given the ambiguous history of religion in every culture, this makes some sense. And given that many of the best religious thinkers have confined their attention to clarifying the religious vision of a particular tradition for the sake of that tradition alone, this reluctance to discuss religion in the public realm also makes some sense.

However, if we are to find out if we share any basic values (visions of the good) at all about what is human about a human being, the religions can and should be viewed as traditions of great and subtle complexity on these very issues and, at their best, as ancient and highly developed depositories of rare wisdom for any open-minded inquirer. If we are to hold, for example, that any responsible understanding of human being precisely as human must include some acknowledgment of our embodied, relational, and responsible character as human beings, then we may well want to consider the following examples from religious traditions.

1. On the necessary embodiment of the human being as a self, there are few wiser traditions (even the Aristotelian) than the Jewish tradition in its extraordinary and unbroken defense of the reality of human embodiment for authentic humanity. The book of Genesis alone, for example, is clearly as wise a text as our culture possesses on how human beings precisely as humans are embodied, and how they cannot be viewed as merely autonomous minds and wills. The Rabbinic, Kabalistic, and contemporary Jewish reflection—Reform, Conservative, and Orthodox—bears such subtle and persuasive analysis of the reality of eros as embodied, of mind as embodied, of will as embodied in the innate bonds constituting a people, that we ignore these classic Jewish discussions at the price of impoverishing ourselves as a society. Among the religions, possibly only Taoism provides so rich a resource for reflection of the full implications of our embodiment for our humanity as Judaism does.

2. On the intrinsic relationality of every person, there are a few wiser, centuries-long traditions of reflections than the Catholic social justice tradition. In Catholic theory, there is no concept of the modern possessive individual, even in the recent and strong Catholic defense of individual human rights (especially in Pope John Paul II's writings). The central Catholic category for the human is usually the dignity of the person—and the "person" is always understood as an intrinsically relational reality. There is no such reality, for Catholic reflection, as a human person without an intrinsic relationship to other persons, to the community, to nature, and to god. Nor, as John Courtney Murray argued, was there for the American founders and their appeals to "self-evident truths." Once again, to ignore the complexity and subtlety of arguments in the Catholic tradition of reflection on the reality of the human per-

son as relational and the unreality of the "possessive individual" is to impoverish the public discussion on the nature of human beings as human.

3. In the Reformation traditions, moreover, the classic "Protestant principle" of critique and suspicion is strong. Where else in the history of the religions can one find such useful reflections on the self's propulsion to self-delusion? Where else does one find contemporary ethical reflection like Paul Ramsey's or Reinhold Niebuhr's or James Gustafson's on how technology and science, however necessary and admirable, never remove such basic human drives as power, pride, and greed? Indeed, this prophetic principle and its explicit appeal to some notion of a responsible self is central to all prophetic traditions. Recall for example, Martin Luther King, Jr.'s brilliant use of biblical motifs for a genuinely public discussion of the just and loving society. This prophetic vision of justice lives in all the great monotheistic traditions from the prophets of the Hebrew Bible through the Reformers of Christianity to the amazing single-mindedness and purity of will of so many Islamic traditions. Today we need the prophetic principle, above all, to keep reminding ourselves, as the classic prophets always did, that justice for the poor, the oppressed, the marginal of society is the true moral test of the genuine civilization of that society. If human cloning becomes (as it easily and unintentionally could without constant vigilance and reflection—and prophetic outcry) merely another luxury item for the powerful, the rich, the talented, the beautiful, we have, as Amos, Isaiah, and Jeremiah would not hesitate to say, damned ourselves as a people.

As such thinkers as Franklin Gamwell have argued with care, moreover, the question of God can and should become, on commonly available grounds of public reason, a question for any inquirer in the public realm. But even before that further question of God is addressed publicly, the resources of the religions on understanding shared intuitions on the embodied, relational, justice-driven character of human beings as human beings can and should be part of the wider public debate on human cloning. Of course the use of the Western monotheistic religions' visions of the human good should not exclude but encourage inquiry into alternative visions of the human good in the tragic visions of the West, in the great philosophies and works of art—popular and elite—of the ancient, medieval, and modern periods. All are public resources. It is weirdly self-impoverishing to ignore that cultural fact.

Of course, we must also be open to learn from other religious traditions as well on the human good: the remarkable insights of the Taoist traditions, especially on the body; the unparalleled wisdom of the Buddhist traditions, especially on our relationships to non-human creatures and our need to cease clinging to our possessive egos; the clarity of the Confucian tradition and its exceptional insight into our responsibilities to past and future generations; the rich complexity of the Hindu traditions on the reality of the erotic in all spiritual quests for humanity; the wisdom of such indigenous traditions as our own native American spiritual traditions on our human selves in community not only with our fellow humans but also with nature and the cosmos.

These examples of resources from the religions are so briefly stated here, I admit, that they may, at the moment, seem more like "hints and guesses" than the needed lengthy description and defense of their intuitions and visions of the human good of the religious traditions. But like any suggestive examples, they may at least serve to remind us of some of the resources we do, in fact, already possess, if we are wise enough to employ them. Fortunately, on the debate on the possibility of human cloning we are not yet at what too often passes as public debate on visions of the good: shouting matches masquerading as debates; ever more clever marketing devices for new consumer goods (including cloning?); scientific, philosophical, artistic, and religious monologues unwilling to hear one another. At least bumper stickers have not yet replaced reflection on the possibility of human cloning.

There is still time for the communal discussion to be demanding, clear, and inclusive. Ultimately no one of us will be able to avoid the literally awe-full questions that the possibility of human cloning provokes for any thoughtful person: What do we ultimately mean by a human being? For myself, I remain profoundly suspicious of human cloning as even a possibly positive contribution to the human good. But everything I know and sense about my own relative ignorance on certain aspects (and not only technical scientific ones, but also the many ethical issues relevant to this debate) also leads me to acknowledge my own need to listen, hear, and learn from others. Surely I am not alone in this sensed need. If we cannot discuss reasonably and openly the unavoidable question evoked by the possibility of human cloning—"What is a human being?"—then we might as well all fold up our tents and return to whatever private reservation of the spirit we inhabit. For then we would have to admit that there is no genuine public realm where all can and must meet on those questions that necessarily involve us all.

PART IV
Law and Public Policy

The Constitution and the Clone

Cass R. Sunstein

"My decision to clone myself should not be the
government's business, or Cardinal O'Connor's, any more
than a woman's decision to have an abortion is. Cloning is
hugely significant. It's part of the reproductive rights
of every human being."

—RANDOLFE WICKER, head of the Cloning Rights United Front
of New York, *New York Times Magazine*, May 25, 1997, at 18.

*I*t is some time in the future. Federal and state laws forbid the practice of
cloning human beings; but a number of American citizens are claiming
that the Constitution guarantees a "right to clone." Some infertile couples
contend that cloning provides their "best option for having children." Some
couples who have lost their first child are seeking a biologically identical replacement via cloning. Some homosexual groups are insisting on a right to
clone.

An especially prominent case was brought by Kristina Martin and
Ronald Martin. Kristina is 38 and Ronald is 42; Ronald is infertile.
Kristina seeks to carry Ronald's cloned child. In a case brought by the Martin and several other married couples, a court of appeals held that the Constitution protects the right to clone. The constitutional issue has come before
the Supreme Court of the United States.

But now there is a fork in the road: two possible but very different paths for constitutional law. Path A contains the Supreme Court's reasoning in its dramatic decision firmly recognizing a constitutional right to reproductive privacy, including the right to clone. Path B summarizes the Supreme Court's reasoning in its unambiguous denial of any such right. There are intersections between the two divergent paths. The reasoning of the two Supreme Courts has been organized to allow comparisons between the two paths—and to facilitate choices between them.

Kristina Martin and Ronald Martin, et al. v. Martin Ballinger, Secretary of Health and Human Services, et al.

On Petition for Writ of Certiorari to the United States Court of Appeals for the Eighth Circuit

No. 99-1099.

Justice Monroe delivered the opinion of the Court.

In this case American citizens seeking to clone human beings have challenged federal and state prohibitions on cloning. The plaintiffs are married couples. They argue that these prohibitions violate their right to free reproductive choice.

The plaintiffs make two arguments. They argue, first, that the right to clone is part and parcel of the right of reproductive privacy. They claim that this right is akin to the rights to use contraception and to have an abortion, firmly recognized under our prior cases. They therefore urge that the right to clone qualifies, under the due process clause of the Fifth and Fourteenth Amendments, as a "fundamental interest," which may be invaded only if the government can satisfy "strict scrutiny," by showing the most compelling of justifications.

The plaintiffs argue, second, that even if the right to clone does not qualify as a "fundamental interest," the legal prohibition on cloning is unconstitutional, because the government cannot show a "rational basis" for the restriction. The plaintiffs claim that the government has no legitimate reason for interfering with private decisions about whether to have a child via cloning.

The government makes several arguments in response. It urges that the right to clone is very different from the right to use contraceptives or to have an abortion, and that it does not qualify as a fundamental right under the Constitution. The government also claims that there are compelling reasons to ban human cloning even if the right to clone does qualify as a fundamental right. Finally, the government insists that there is a "rational basis" for restricting human cloning, to prevent a wide range of social harms.

We reject the government's arguments and hold today that under the Constitution as it has come to be understood, there is a constitutional right to clone. The Court's cases firmly recognize the individual right to control reproduction—to decide whether or not to have a child. This right lies at the very heart of the constitutional right of "privacy"; it is part of the liberty protected by the due process clause of the Fifth and Fourteenth Amendments.

To be sure, the government can override that right if it has an extremely good reason for doing so. But in the context of cloning, the government's arguments are far too weak. Certainly the government cannot satisfy the "strict scrutiny" standard that governs this Court's review of restrictions on fundamental rights. Indeed, we do not believe that the government's interests are strong enough to overcome "rational basis" review, the most deferential standard the Court now uses.

I

There is an acknowledged constitutional right to some form of individual control over decisions involving reproduction. In *Griswold v. Connecticut,* 381 U.S. 479 (1965), this Court held that a state could not ban a married couple from using contraceptives. In the

Court's view, a right of "privacy" forecloses state interference with that decision. In *Eisenstadt v. Baird,* 405 U.S. 438 (1972), the Court extended *Griswold* to invalidate a law forbidding the distribution of contraceptives to unmarried people. In the key passage of its opinion, the Court said, "If the right of privacy means anything, it is the right of the individual, married or single, to be free from unwarranted governmental intrusion into matters so affecting a person as the decision whether to bear or beget a child." And in *Roe v. Wade,* 410 U.S. 113 (1973), the Court held that there is a constitutional right to have an abortion. This decision the Court strongly reaffirmed in *Casey v. Planned Parenthood,* 505 U.S. 833 (1992).

On the other hand, we have held that there is no general right against government interference with important private choices. Thus in *Washington v. Glucksberg,* 116 U.S. 2021 (1997), we held that under ordinary circumstances, there is no general right to physician-assisted suicide. We emphasized that this right is quite foreign to the traditions of American law and policy, which strongly discourage suicide, assisted or otherwise.

These cases clearly establish a basic principle: The Constitution creates a presumptive individual right to decide whether and when to reproduce. The precise dimensions of this right will inevitably change over time. Of course new technologies are expanding the methods by which reproduction is possible. New technologies have been especially prominent in the last decades, and undoubtedly scientific progress will continue, producing unforeseeable developments. The Constitution provides the basic right, which is itself constant; but the specific content of the right necessarily changes with relevant technology.

We think it very plain that the government would need an exceptionally powerful justification to ban couples with serious fertility problems from using methods other than sexual intercourse to produce a child between husband and wife. It has become quite ordinary for couples to use new methods and technologies, and governments have generally refrained from interfering with indi-

vidual freedom to choose. A governmental ban on in vitro fertilization, to take one example, would have to be powerfully justified,
certainly if the ban prevented couples from having children in the
only way they could. Cloning is of course a new technology. But for
some couples, including the plaintiffs here, it is the only or the
best reproductive option. If the government wishes to limit or restrict that choice, it must come up with an exceptionally strong justification.

II

The closest precedent for our decision today is *Roe v. Wade,* reaffirmed in *Casey v. Planned Parenthood,* and these decisions strongly
support a right to clone. In fact the argument for a right to clone
is far stronger, in many ways, than the argument for a right to abortion. Cloning produces life where abortion destroys it. Abortion is
contested on the ground that it destroys the fetus, which many
people consider to be equivalent to, or nearly equivalent to, a
human being. If there is a right to abort fetal life, there must be a
parallel right to create life. Of course the morality of abortion, like
that of cloning, is socially contested. *Roe* demonstrates that the fact
that people have moral reservations, whether or not inspired by religion, is by itself an insufficient reason to allow interference with
a decision about whether to bear or beget a child.

In any case the right to have an abortion reflects a judgment
that the choice about whether to reproduce lies with the individual, not the government. That judgment strongly supports the
plaintiffs' claim here.

III

If a ban on cloning is subject to the most stringent forms of constitutional review, it is clear that the ban cannot be upheld. The government's interests are speculative in the extreme. The government

notes that there is a risk of psychological harm to the clone, who will know that it is the genetic equivalent of someone else, with a known life; perhaps this knowledge will be hard to bear. We acknowledge that psychological harm may occur. But psychological harm is a risk in many settings, and it is not a reason to allow the government to control reproductive choices, to ban adoption, or (for that matter) to outlaw twins, for whom there is in any case no decisive evidence of trauma.

The government argues that it fears the outcomes of unsuccessful medical experiments; it says that children with various physical defects are likely to occur. This too is possible, especially at the early stages, but it is a reason for regulation, not for prohibition. The government refers as well to the need for a large stock of genetic diversity. The interest in a large gene pool is, we may acknowledge, compelling; but it is utterly implausible to think that the existence of cloning, bound to be a relatively unusual practice, will compromise the genetic diversity of mankind.

Finally, the government attempts to justify its ban with the legal equivalent of tales from science fiction or horror movies—thus, the government refers to dozens or even hundreds of genetically equivalent people, or of clones of especially abhorrent historical figures. The government fears that narcissistic or ill-motivated people will produce armies of "selves." We think it plain that these fanciful speculations, far afield from the case at hand, do not justify a total ban. If problems of this kind arise, they should be controllable through more fine-tuned regulations. The government's emphasis on unlikely scenarios of this kind simply confirms our belief that the government has been unable to find a "compelling" interest to override the presumptive right to control one's reproductive processes.

IV

Even if the right to clone did not qualify as a fundamental interest, we believe that a wholesale ban on cloning would be unconstitutional, because it cannot survive rational basis review. In the end the government's justifications are best understood as a form of unmediated and highly emotional repugnance—produced not by evidence, arguments, or reality, but by the simple novelty of the practice under review.

Repugnance frequently accompanies new technological developments; and repugnance tends to underlie the worst forms of prejudice and irrationality. We need not repeat the details about our nation's long-standing practices of discrimination on the basis of race and sex—now understood to violate our deepest constitutional ideals—in order to establish the point. Nor should it be necessary to stress that the most solemn obligation of this Court is to uphold constitutional principles against popular prejudice and irrationality, which often take the form of "repugnance." And to the extent that the ban on cloning has foundations in religious convictions, it should be unnecessary to say that religious convictions, standing alone, are not, in a pluralistic society, a sufficient basis for the coercion of law. It does not deprecate religious conviction to say that its appropriate place is not in the statute books, and to emphasize that religious arguments must have secular equivalents in order to provide the basis for law.

V

Our conclusions are fortified by a simple, widely recognized point: A ban on cloning will simply drive the practice of human cloning both abroad and underground. At this stage in our history, it is altogether clear that prohibitions on cloning will not operate as prohibitions, but will simply force people who are determined to clone to act unlawfully or in other nations. Thus, the prohibition at issue here cannot be supported by the government's justifications, which would lose what little force they have if the prohibition cannot operate in practice as it does on paper.

We hold, in sum, that the right to clone, however novel as a matter of technology, is part and parcel of a time-honored individual right to control the circumstances and event of reproduction; that the government has pointed to no sufficient justification for overriding that fundamental right; and that on inspection, the government's grounds for concern dissolve into a simple statement of repugnance and disgust, lacking scientific or ethical foundations and fed mostly by imaginative literature. Because they invade the right to privacy in its most fundamental form, legal bans on cloning violate the due process clause of the Constitution.

The judgment of the court of appeals is affirmed.

It is so ordered.

Kristina Martin and Ronald Martin, et al. v. Martin Ballinger, Secretary of Health and Human Services, et al.

On Petition for Writ of Certiorari to the United States Court of Appeals for the Eighth Circuit

No. 99-1099.

Justice Winston delivered the opinion of the Court.

In this case a group of American citizens seeking to clone human beings have challenged federal and state prohibitions on cloning. The plaintiffs are married couples. They argue that these prohibitions violate their right to free reproductive choice.

The plaintiffs make two arguments. They argue, first, that the right to clone is part and parcel of the right to reproductive privacy, closely akin to the rights to use contraception and to have an abortion, recognized under our prior cases. They therefore urge that the right to clone qualifies, under the due process clause of the Fifth and Fourteenth Amendments, as a "fundamental interest," which may be invaded only if the government can satisfy "strict scrutiny," by showing the most compelling of justifications.

The plaintiffs argue, second, that even if the right to clone does not qualify as a "fundamental interest," the legal prohibition is unconstitutional, because the government cannot show a "rational basis" for the restriction. The plaintiffs claim that the government has no legitimate reason for interfering with private decisions about whether to have a child via cloning.

The government makes several arguments in response. It urges that the right to clone is very different from the right to use contraceptives or to have an abortion, and that it does not qualify as a fundamental right under the Constitution. The government also claims that there are compelling reasons to ban human cloning even if the right to clone does qualify as a fundamental right. Finally, the government insists that there is a "rational basis" for restricting human cloning, to prevent a wide range of social harms.

We accept the government's arguments and hold today that under the Constitution as it is now understood, there is no constitutional right to clone. The Court's cases recognize an individual right to control reproduction, as part of the liberty protected by the due process clause of the Fifth and Fourteenth Amendments. But it is facetious, at best, to say that anything in our precedents recognizes an individual right to replicate oneself through the new technology of cloning.

To override the individual interest in replicating other human beings, the government needs only a "rational basis." But the government has far more than this. We believe that the government has exceptionally powerful grounds for controlling cloning. Indeed, the government's justifications are strong enough to override the individual's interest even if the government needs to overcome "strict scrutiny," the least deferential standard the Court now uses.

I

There is an acknowledged constitutional right to some form of individual control over decisions involving reproduction. In *Griswold v. Connecticut,* 381 U.S. 479 (1965), this Court held that a state could not ban a married couple from using contraceptives. In the Court's view, a right of "privacy" forecloses state interference with that decision. In *Eisenstadt v. Baird,* 405 U.S. 438 (1972), the Court extended *Griswold* to invalidate a law forbidding the distribution of contraceptives to unmarried people. In the key passage of its opinion, the Court said, "If the right of privacy means anything, it is the right of the individual, married or single, to be free from unwarranted governmental intrusion into matters so affecting a person as the decision whether to bear or beget a child." And in *Roe v. Wade,* 410 U.S. 113 (1973), the Court held that there is a constitutional right to have an abortion. This decision the Court strongly reaffirmed in *Casey v. Planned Parenthood,* 505 U.S. 833 (1992).

On the other hand, we have held that there is no general right against government interference with important private choices. Thus in *Washington v. Glucksberg,* 116 U.S. 2021 (1997), we held that under ordinary circumstances, there is no general right to physician-assisted suicide. We emphasized that this right is quite foreign to the traditions of American law and policy, which strongly discourage suicide, assisted or otherwise.

These cases clearly establish a basic principle: The Constitution creates a presumptive individual right to decide whether and when to reproduce. Thus, government cannot prevent people from choosing not to have a child, and we agree with the plaintiffs that serious issues would also be raised by (for example) a legal requirement of abortion, or a restriction on the number of children a married couple might have. But the constitutional right is far from unbounded. It lies in a specific judgment about *reproduction,* understood by our traditions as a distinctive human interest with a distinctive human meaning.

Our traditions rebel against the idea that the state, rather than the individual, can make the decision whether a person is to bear or beget a child. But it defies common sense to suggest that *repli-*

cation falls in the same category. No tradition supports a right to replicate. This is not merely a matter of technological limitations. The human meaning of replication, of creating genetically identical beings, is fundamentally different from that of reproduction, and replication is to many people horrifying. Centuries of culture, of myth and literature, confirm this basic fact. It is not the business of this Court to say whether replication of human beings should or should not be permitted. But when the people of the country, and their elected representatives, conclude that it should be banned, no fundamental right is invaded. We require only a rational justification.

II

The plaintiffs rely most fundamentally on *Roe v. Wade,* but there is an enormous difference between the right to clone and the right to an abortion. The Court has come to see that the decision in *Roe* turned in large part on the interest in equality on the basis of sex. As a matter of history, governments have denied the right to abortion because they seek to preserve women's traditional role. Moreover, the denial of the right to have an abortion tends to fortify that traditional role and thus to undermine equality on the basis of sex. No equality interest supports the right to clone. With respect to sex equality, cloning is a very complex matter—reasonable people have set forth competing views—and we do not believe that it is plausible to argue that the right to clone finds a justification in principles of equality on the basis of sex.

There is a further point. *Roe v. Wade* did not create a general right to decide whether to have a child through whatever technological means may be available. It is far narrower than that: an outgrowth of cases establishing a right to decide whether to reproduce, a time-honored right in Anglo-American law. The right to replicate stands on much weaker ground.

III

Even if the ban on cloning were subject to the most stringent forms of judicial review, it would be upheld. Physical difficulties and even deformities are highly likely. The government has an exceptionally strong interest in protecting young children against disease and disability, and both of these are likely products of experiments in human cloning. The government has pointed to considerable evidence of this risk, and this Court is in no position to second-guess the scientific evidence.

It is highly likely as well that the practice of cloning would have undesirable effects on the "people" who result. There is some evidence of psychological difficulties faced by human twins; the practice of cloning will inevitably risk far more severe problems from people who know that they are genetic equivalents of people with known lives, including known problems, known successes, and known failures. Similarly, the government reasonably fears that the parent-child relationship, between genetic equivalents, would be unrecognizable, and permeated by difficulties of various sorts. It may well be that especially wealthy people, or especially narcissistic people, would fund large numbers of replications of themselves. Who would rear the resulting children? With what motivations? The government has the strongest possible reasons to fear the outcomes of such a situation.

There is a further consideration. The government has an extremely powerful interest in preserving the stock of biological diversity. Widespread cloning could compromise that interest to the detriment of humanity as a whole. We think that these points confirm our belief that the government has a wide range of compelling interests sufficient to override any presumptive right to clone.

IV

If the decision whether to clone does not qualify as a fundamental right, the only question is whether the government's ban is "rational." Certainly it qualifies as such. As we have indicated, the government might reasonably believe that there would be adverse psychological effects on "clones." It might believe that the process of cloning human beings would result in physical deformities for the "products" of the relevant scientific practices. It might even believe that the existence of clones would have adverse psychological effects on many children and even adults who might fear that they would be cloned against their will. All of these speculations are reasonable.

This Court does not sit to second-guess reasonable judgments by the elected representatives of the American people, especially when there is no defect in the system of democratic deliberation that gave rise to the law under review. And it should not be necessary to say that the religious convictions that may underlie legal bans on cloning are not, in a pluralistic society, at all troublesome from the standpoint of constitutional democracy, where everyone's convictions are entitled to count.

Our conclusions are not undermined in the least by the possibility that people determined to defy the law will do so, either here or by seeking refuge abroad. It is not an argument against the criminal law that criminal prohibitions may be violated or circumvented. If bans on human replication do not operate in practice as they do on paper, Congress and the legislatures of the several states are entitled to respond as they choose.

We hold, in sum, that the right to clone has no basis in our constitutional traditions, which involve human reproduction, not replication; that the government has ample grounds for restricting individual choice in light of the novelty of the relevant technology and its unpredictable and potentially damaging effects on the most basic human values and in particular on those involved in the relevant "experiments." Because they do not implicate the right to privacy or any other constitutionally protected interest, legal bans on cloning are entirely consistent with the due process clause of the Constitution.

The judgment of the court of appeals is reversed.

It is so ordered.

On Not Banning Cloning for the Wrong Reasons

Laurence Tribe

Most of the interesting issues about cloning in general, and human cloning in particular, involve shades of gray—not "either/or" but "how": How should the evolution and deployment of this fascinating but disturbing new technology be funded and controlled? How should the human experimentation that would precede its routine availability for the genetic replication of people as opposed to sheep or calves be conducted and regulated? How should domestic and international regulatory efforts mesh? Interesting though such questions are, they are not my focus here. Rather, I want to address the strictly binary question: Should human cloning be forbidden altogether? Although I offer no definitive answer to that inquiry, I explore what I have come to regard as a particularly forceful objection to any regime of prohibition in this technological realm—an objection centered on how the very *fact* of prohibition, and the social meanings likely to attend it, might be expected to reshape, in strikingly negative ways, the structure of social relations and the status of the human lives that those social relations in turn affect.

As it happens, this isn't my first attempt to address these questions. The first time around, I had only recently completed law school and a pair of judicial clerkships and had just begun my career as a teacher of law building on a one-year stint directing a study of technology assessment for the National Academy of Sciences. Maybe a bit too eager to use that technology-focused experience as a springboard for my musings about emerging advances in biomedical science and its applications, I seem to have been predisposed to view significant technological changes in the most portentious terms—and to pay somewhat less attention to the social and cultural impacts of the legal regimes through which new technologies are either facilitated or forbidden. Now, after a long period of reflection about the structure of law generally, and of constitutional law in particular, I guess I'm more attuned to the effects of legal rules themselves—and of legal prohibition as an especially consequential "technology" in its own terms. And so it is, as the brief essay that follows will hopefully make plain, that I'm a lot more dubious now than when I first explored this terrain about the wisdom of banning the cloning of human beings.

In my first crack at this subject—taken at a time when most people didn't take the prospect of human (or, indeed, of mammalian) cloning seriously enough to warrant careful analysis—my thinking took the form of a series of articles about the need to assess and channel new biomedical and electronic technologies not solely in instrumental terms—that is, not solely in terms of their "costs" and "benefits" as measured on existing scales of value—but in intrinsic terms as well. In those articles, I urged the importance of assessing at least the most basic new technologies, especially those operating directly on the human mind or body, in terms of how their development and use might alter "the ends—and indeed the basic character—of the individuals and the communities that choose them."[1] The articles focused in particular on what struck me as the paradigm case of a technology requiring assessment in such intrinsic terms—a reproductive method that seemed to lie not very far ahead: the "perfection of cloning technology for human beings."[2] My evaluation of that reproductive technology stressed the fragility of purely consequentialist arguments against such a development.[3] Any "on balance" assessment against allowing at least the selective use of such a technology, grounded in the estimate that any such use would inflict greater cost than benefit in terms of "our" values, would be vulnerable to potent rebuttal—(1) in terms of the necessarily inconclusive nature of any such estimate; (2) in terms of the question-begging character of appeals to shared values where how "we" should value various outcomes is anything *but* shared; and (3) in terms of the doubtfulness of assuming that the choice of how an individual or a couple should be permitted to reproduce is properly to be made through collective political processes. If we ask whether the concrete gain to the individual or couple and their child-to-be is outweighed by the less tangible loss to their community and perhaps to future generations, we may be hard-pressed to explain why that balance is the state's rather than the individual's to strike. And alleged losses to the child in question are particularly tough to factor into the cost/benefit calculus when the alternative to that child's clonal creation is something as stark as nonexistence. How awful to be a clone? How awful not to *be* at all?

In those early writings, without ever coming out four-square for a ban on human cloning, I suggested that the strongest case for nipping human cloning in the bud, or for banning its use on a global basis once such use became feasible (and perhaps for banning particular forms of research and development thought likely to lead in an imminent way to the feasibility of human cloning, although efforts to forbid the generation of supposedly dangerous knowledge might well be deemed intolerable in terms of freedom of inquiry), would probably take an intrinsic rather than instrumental form. My argument was that cloning human beings might well constitute "a fundamental threat to the concept and the reality of the human person as a unique and intrinsically valuable entity, conscious of its own being and responsible for its own choices."[4] The essay quoted the physician-philosopher Leon Kass, expressing agreement with his suggestion that "to 'lay one's hands on human generation is to take a major step toward making man himself simply another one of the manmade things.' "[5] And the essay went further. To conceive of the changes that would be wrought by permitting such a technology to be deployed "as a selection in terms of a 'given' value framework," I claimed, "begs the question presented" by the technology's use, for "[a]t stake are not merely alterations in the 'costs' and 'benefits' associated with implementing existing preferences and values but alterations in the very structures of human thought and reality on which all value premises and the choices that embody them—all the frames of reference for defining one thing as a 'cost' and another as a 'benefit'—must ultimately be based."[6] Many others before and since—with ideological orientations as different as those of Karl Marx[7] and Leon Kass[8]—have likewise made much of how every major choice among technologies is likely to reconstitute the choosers, both as developers and as users, in often profound ways.

This recap of my earlier work reflects no sense of self-congratulation, but a spirit of confession. For the truth is that, twenty-five years after first seeking to apply to the example of human cloning the insight (by no means original with me) that certain technologies might reconstitute society too deeply to yield to merely instrumental analysis, my primary impulse is to say that my "yes" answer to the truly basic question of whether human cloning must be perceived as a threat to the very meaning of human individuality was probably wrong. No doubt human cloning is deeply unsettling to many, perhaps most, of us. But would its advent jeopardize something truly essential? My answer today is: maybe—but maybe not. Who was I—who is anyone—to forecast *which* technologies will over time generate transformations sufficiently deep, and sufficiently worthy of shared condemnation, that we may confidently favor the outright prohibition of those technologies, predicating such prohibition on our prediction that the technologies we urge banning would otherwise deform the human project?

It's certainly not a matter of technological manifest destiny—not a matter, that is, of assuming that anything people are technically capable of doing ought, for that reason alone, to be permitted. It's not even a matter of positing a basic human right—or, in American law, a federal constitutional right—to reproduce oneself through whatever means become technically possible. Whether courts should declare such a right, and what its limits might be, is a subject for another essay. It's simply a matter of humility: My point is less to make a federal case out of it than to suggest a far more tentative attitude. How can any of us feel so confident that the meaning of humanity will be degraded by human cloning in any and all circumstances that we are prepared to shut down altogether the potentially humane possibilities of this admittedly (for most of us) most distasteful, perhaps even diabolical, method of reproducing human life?

Perhaps some technological possibilities—those that would significantly alter the genetic or at least biological composition of the human species as a whole, or those that would "marry" humanity with its humanly developed systems for processing and exchanging information so as to create human-computer hybrids that might have to be regarded as new species—would genuinely reshape humanity itself, and would accordingly require all who debate the pros and cons of the underlying choices to grapple with the truly ultimate questions of good and evil posed by proposed changes in "human nature." Whether to permit the genetic blending, for experimental or other purposes, of human beings with chimps or with computer chips—whether to allow the generation of chimeras or of cyborgs—might pose questions of this deep character. Does human cloning pose those questions? Some of the thinkers with whose work on this topic I agreed in the 1970s argue that it does; they have reacted to the now far more imminent prospect of human cloning essentially by reiterating and elaborating their earlier convictions as to the nightmarish, nature-altering character of such a technology.[9] Although I continue to think those misgivings point to real problems and to issues far too profound to be ignored, I have come to believe that objections of the sort that writers like Kass put forth—objections that rest ultimately on an aversion to patterns of human interaction and behavior that the observer deems "unnatural"[10]—are themselves at *least* as serious a source of danger as are the technologies that provoke them. We *may* have more to fear than fear itself—but when fear of the unnatural drives a campaign to ban some innovation, then that very fear may be more fearsome than the innovation that spawns it.

It's not that the amorphous and often hard-to-articulate character of such fears or objections is *itself* proof of their misguided character. Wisdom often outpaces our ability to capture its essence in verbal formulas, and those who automatically dismiss deeply felt misgivings as insubstantial or as irrationally sentimental whenever we have not (yet) been able to capture those misgivings in a rigorous argument underestimate the profundity of human intuition—and overestimate the power of cold logic. On the other hand, the difficulty of capturing in crisp and tightly analytical form the visceral unease one might feel at the deployment of a given technology should not be confused with a sign that this unease *must* be well-grounded, that it must reflect a wisdom too deep for words. Unease should count for a great deal; but it must not count for everything.

When that unease focuses on the sense that a proposed act or technical development is incompatible with what the opponent of that act or development deems the essence of human nature or of the human condition, the alarm bells ought to go off loud and clear. It is no coincidence, it seems to me, that the latest eloquent assault on cloning by Leon Kass is embedded in an essay that describes "[c]loning" as "the perfect embodiment of the ruling opinions of our new age," in which "the sexual revolution" has made it possible for us to "deny in practice, and increasingly in thought, the inherent procreative teleology of sexuality itself," in which, once "sex has no intrinsic connection to generating babies, babies need have no necessary connection to sex," and in which, "[t]hanks to feminism and the gay rights movement, we are increasingly encouraged to treat the natural heterosexual difference and its preeminence as a matter of 'cultural construction.' "[11] The Kass crescendo ends with this observation: "Thanks to the prominence and the acceptability of divorce and out-of-wedlock births, stable, monogamous marriage as the ideal home for procreation is no longer the agreed-upon cultural norm. For this new dispensation, the clone is the ideal emblem: the ultimate 'single-parent child.' "[12] So there we have it: Cloning is the technological apotheosis of Murphy Brown and Ellen DeGeneres, the biomedical nemesis of Dan Quayle, Phyllis Schlafly, and Pat Robertson.

I say that this linkage—between passionate opposition to human cloning as a lawful option, and emotion-charged objection to supposedly "unnatural" ways of living and loving and parenting—is no coincidence precisely because the cloning objection being addressed here, like the objections to such practices as surrogate motherhood or gay marriage and gay adoption, takes the form of an irreducible appeal to human nature, whether or not divinely ordained, as the normative source of the case for legal prohibition. Much has, of course, been written about the virtues and vices of such appeals in other social and moral contexts. Here, my focus is on the special problems inherent in claims that a particular method of *creating human beings* is unacceptably unnatural and hence ought to be forbidden as evil. (Let's put to one side objections to various courses of action, presumably including the creation of embryos designed to be discarded, that might have to be undertaken as part of any research and development program leading to the perfection of the reproductive method in question; a case for banning such embryo-manipulating behavior on the ground that it constitutes unethical experimentation on human beings without their consent might well be convincingly advanced quite apart from the supposed case against any use of the particular reproductive method whose perfection the course of experimentation might bring about. Once the contested reproductive technique has finally been developed—outside the United States, perhaps—objections to the experiments that were needed to produce it are likely to be irrelevant, leaving us with the core question of whether the mode of human reproduction, having been developed, ought to be made criminal.)

As we analyze any proposed decision to outlaw a particular way of making babies, it's vital to keep in mind that, just as physical or biomedical technologies may in some circumstances "reconstitute" the societies that choose them by altering the preferences and even the values of those societies, so too there are significant constitutive dimensions in any society's decisions to accept a line of argument against deploying a given technology, and then to employ the institutional "technology" of domestic legal prohibition (presumably coupled with international agreements designed to make the prohibition meaningfully enforceable) in an effort to prevent the technology in question from ever being used. It is not only *constitutional* law but *all* of law that has important constitutive effects, and those effects may be particularly dramatic when the technology of law—including the legal practices, arguments, and institutions that support and surround a decision to criminalize a particular physical or biomedical technology—is deployed in an effort to prevent that physical or biomedical technology from being used. Even when that contested technology would operate to do something other than create a human child—when it operates, for instance, to abort a developing embryo (as with RU 486, say) or to induce addiction to a pleasurable substance (as with the nicotine delivery system of a cigarette)—employing the counter-technology of criminalization may well have, as all are aware, the many "costs" associated with black markets generally. Whether one thinks of coat-hanger, back-alley abortions before 1973,[13] or of the concern with how an FDA ban on nicotine in cigarettes might encourage addicts to resort to a contraband market, a standard part of the policy analysis surrounding choices between legalization and criminalization entails addressing the consequences of imperfect (often grossly imperfect) enforceability.

When the technology at issue is *a method for making human babies*—whether that method differs from a society's conventional

and traditionally approved mode because of some socially con-
structed "fact" such as the marital status or kinship relation or racial
identity of a participant, or differs in a more intrinsic way as in the
case of in vitro fertilization, or surrogate gestation, or cloning so
as to achieve asexual reproduction with but a single parent—
applying the counter-technology of criminalization has at least one
additional, and qualitatively distinct, social cost. That cost, to the
degree any ban on using a given mode of baby making is bound to
be evaded, is the very considerable one of creating a class of potential outcasts—persons whose very *existence* the society has chosen, through its legal system, to label as a misfortune and, in
essence, to condemn.

Even the simple example of what the "politically correct" call
nonmarital children and what others call illegitimates (or, more
bluntly, bastards) powerfully illustrates the high price many individuals and their families are forced to pay for a society's decision
to reinforce, through outlawing nonmarital reproduction and dis-
criminating against nonmarital offspring, particular norms about
how children ought to be brought into the world. How much
higher would that price be when the basis on which the law decides
to condemn a given baby-making method (like cloning) is not simply a judgment that the particular pairing of parents was for some
reason to be frowned upon to the point of prohibition, but the far
more personalized and stigmatizing judgment that *the baby itself*—
the child that will result from the condemned method—is morally
incomplete or existentially flawed by virtue of its unnaturally man-
made and deliberately determined (as opposed to "open") origin
and character? That judgment may begin with sympathy or even
pity for the innocent baby's anticipated predicament as a child
whose single parent is the baby's identical twin and whose very
being may express and embody parental expectations with which
no child should be burdened. But even if the judgment begins with
sympathy, its very structure entails an enormous risk that it will end
in condemnation, in some version of original sin. At first the ob-

ject of a pity that may be justified (but might also be misplaced—
are not many *non*-clones even more cruelly burdened by parental
expectations?), the human clone—in a world where cloning is forbidden as unnatural—is likely in the end to become the object of
a form of contempt: the contempt that the (supposedly) spontaneous, natural, and unplanned would tend to feel toward the (supposedly) manufactured and allegedly artificial.

Laws against highly personal activities, conducted in private and
involving no "victim" in the usual sense—sexual prostitution, for
instance—generate, as we have noted, black markets in which such
laws may be circumvented, in turn generating contraband sub-
stances or practices. That is bad enough—though not necessarily a
decisive reason to opt for legalization in any given context, from the
use of narcotic drugs to the sale of body parts. What is far worse
is any situation in which the "contraband" takes the form of human
beings. In any such situation, the upshot may be a particularly per-
nicious form of caste system, in which an entire category of persons, while perhaps not labeled untouchable, is marginalized as
not fully human.

Even if one could (unrealistically) posit perfect enforceability of
a worldwide ban on human cloning, so that the set of contraband
persons would be empty, the social costs of prohibition—whatever
one may think of claims that there is an individual "right" to reproduce by this or, indeed, any other particular method—would be
far from zero. For the arguments supporting the iron-clad prohibition of human cloning we have hypothesized could not easily be
cabined within that single context. Such arguments would almost
invariably rest on, and if acted upon would reinforce, the notion
that it is unnatural, and intrinsically wrong, to sever the conventional links between heterosexual unions sanctified by tradition
and the creation and upbringing of new life. Witness the argument
by Leon Kass, only the most eloquent of a familiar genre. And the
entrenchment of that essentialist notion, its deeper embedding in
our culture and our law, is in turn anything but costless. It is most

assuredly not costless for lesbians, gay men, persons gay or straight with genetically transmittable diseases, and others whose sexual or other orientations or capacities draw them into unconventional patterns of intimate relationship—and unconventional modes of linking erotic attachment, romantic commitment, genetic replication, gestational mothering, and the joys and responsibilities of parenting.

Nor is the entrenchment of the essentialist vision costless for the culture of the wider community, straight as well as gay: A society that bans acts of human creation that reflect unconventional sex roles or parenting models (surrogate motherhood, in vitro fertilization, artificial insemination, and the like) for no better reason than that such acts dare to defy "nature" and tradition (and to risk adding to life's complexity) is a society that risks cutting itself off from vital experimentation and risks sterilizing a significant part of its capacity to grow. This is not necessarily a conclusive argument against banning human cloning, nor does it purport to be. But it is an argument that I hope creates something like a *presumption* against any such ban—an argument one must reckon with and ultimately overcome if such a ban is to be given serious consideration.

In sum, much as one might deplore as shallow and perhaps disgusting the impulse to create genetic "copies" of deceased loved ones—or, maybe even worse, of oneself or of one's heroes—one must be at *least* as suspicious of the impulse to forbid such cloning altogether—and of the legal apparatus through which that impulse would be effectuated—when that impulse is embedded in a picture of human nature, and of human possibility, that is, in the end, neither altogether human nor genuinely humane.

Endnotes

1. L. Tribe, "Technology Assessment and the Fourth Discontinuity: The Limits of Instrumental Rationality," 46 *So. Cal. L. Rev.* 617, 642 (1973) (hereinafter, "The

Fourth Discontinuity"); see also L. Tribe, "Ways Not To Think About Plastic Trees: New Foundations for Environmental Law," 83 *Yale L. J.* 1315 (1974); L. Tribe, "Policy Science: Analysis or Ideology?" 1 *Philosophy & Public Affairs* 66 (Fall 1972); L. Tribe, "Legal Frameworks for the Assessment and Control of Technology," 4 *Minerva* 243, 254–55 (1971).

2. "The Fourth Discontinuity" at 643.

3. *Id.*

4. *Id.* at 648.

5. *Id.* at 649, quoting L. Kass, "Making Babies—The New Biology and the 'Old' Morality," 26 *Public Interest* 18, 49 (Winter 1972).

6. "The Fourth Discontinuity" at 650. For a similar theme in constitutional law and theory, see L. Tribe, "Constitutional Calculus: Equal Justice or Economic Inefficiency," 98 *Harv. L. Rev.* 592 (1985); cf. L. Tribe, "The Curvature of Constitutional Space: What Lawyers Can Learn from Modern Physics," 103 *Harv. L. Rev.* 1 (1989).

7. See "The Fourth Discontinuity" at 650 n. 116.

8. See L. Kass, "The Wisdom of Repugnance," *New Republic* (June 2, 1997), pp. 17–26.

9. E.g., L. Kass, in "The Wisdom of Repugnance," supra note 8.

10. E.g., L. Kass, in "The Wisdom of Repugnance," supra note 8 at 17 ("Dolly was, quite literally, made. She is the work not of nature or nature's God but of man, an Englishman, Ian Wilmut, and his fellow scientists."); *id.* at 20 (considering "the deeper anthropological, social and, indeed, ontological meanings of bringing forth new life . . . cloning shows itself to be a major alteration, indeed, a major violation, of our given nature as embodied, gendered and engendering beings— and of the social relations built on this natural ground"); *id.* at 21 ("A sexual reproduction, which produces 'single-parent' offspring, is a radical departure from the natural human way, confounding all normal understandings of father, mother, sibling, grandparent, etc., and all moral relations tied thereto.").

11. L. Kass, in "The Wisdom of Repugnance," supra note 8 at 18.

12. *Id.*

13. Prior to *Roe v. Wade,* 410 U.S. 113 (1973), the states were free to criminalize nearly all abortions; most did just that.

The Demand for Human Cloning

Eric A. Posner and Richard A. Posner

The news that a sheep ("Dolly") had been created by cloning adult nonreproductive tissue[1] has given rise to speculation that it may soon be feasible to create human beings in the same way. In fact, substantial technical obstacles remain to be overcome,[2] but no doubt they will be in time. The prospect of human cloning is ferociously controversial.[3] The controversy presupposes that if human cloning were safe, reliable, and permitted, there would be a demand for it. For if there would be no demand, why worry? More realistically, if the demand would be slight, or limited to situations that do not provoke acute concern on the part of people who worry about human cloning, there would be no reason to incur the bother and expense of prohibiting it out of fear of monstrous social consequences.

We therefore limit our discussion to the demand for human cloning. We assume that a safe and effective procedure will be developed that enables a man or a woman to produce a perfect genetic copy of himself or herself (or of his or her child—or of anyone, for that matter), a copy that would bear the same genetic relation to the cloned individual that one identical twin bears to the other. We ask, Who will want to take advantage of this procedure, and with what effects: In economic terminology, we focus on the private benefits and social costs of human cloning. We do not consider the demand for cloning in countries in which the demand for children is much greater, and the status of women much lower than in the United States and its peer countries. Nor do we consider the moral and legal issues presented by cloning, such as whether cloning should be permitted without the permission of the person cloned and who would have parental rights over the clone of a person involuntarily cloned. These are not absurd questions; cloning need not be an invasive procedure, since a person sheds many cells every day, any of which might be cloned.

Nor do we attempt to factor into our analysis the sheer "weirdness" of human cloning, a consideration that might be thought to depress the demand. Not only is this consideration analytically intractable, but it is probably only transitional. A product or service that is new and rare tends to be thought of as weird, and its diffusion is resisted. But if it is a source of potentially substantial net benefits, its use will spread, and when some critical mass is reached, the aversion will drop away and a more rapid diffusion will begin.

We are tempted to put to one side the case in which a couple clones its dying child in order to produce a closer replacement than it would get by having another child in the usual way, or in which an infertile couple clones one of the partners in lieu of adoption or (if it is a heterosexual couple and the man is the infertile one) of artificial insemination, or in which cloning is used because one of the partners has a serious genetic disease or weakness. In these situations—situations of "reproductive failure" in a broad sense—cloning might seem to be simply a substitute for the other methods of obtaining a child that do not involve sexual intercourse between the parents. If the demand for human cloning were limited to these situations, the procedure might not seem worthy of greater controversy than in vitro fertilization of long-frozen ova. (Not that modern reproductive technology is uncontroversial; our point is only that human cloning considered merely as an alternative reproductive technology need not raise particularly novel issues.) Yet we shall see later that this may be mistaken—that cloning

the infertile could have the radical consequence of eventually eliminating sexual reproduction. The critical difference between
cloning and other reproductive technologies is not that cloning involves choosing what genes one's child shall have; such choices are
within the horizon of reproductive technology wholly apart from
the Dolly trick of cloning an adult nonreproductive cell. The critical difference is that the other methods require fertility and cloning
does not, or more precisely that cloning does not require that the
biological parent be fertile but only that there be a womb, not necessarily the genetic parent's womb, capable of incubating the clone
embryo.

Since gene selection is not limited to cloning, what we have to
say about the demand for cloning may well have implications for
other reproductive technologies. But we shall generally ignore
those implications. Comparison with in vitro fertilization and the
other now-familiar techniques for overcoming problems of fertility must not be allowed to obscure the fundamental point that the
demand for human cloning would in all likelihood not be limited
to cases of "reproductive failure," broadly construed to include the
child who dies before reaching adulthood and the parent who fears
transmitting a bad gene. The amplification of this point is the main
contribution of this essay.

The principal reason not to expect the demand for human
cloning to be limited to cases of reproductive failure lies in evolutionary biology. A gene's frequency depends on the rate at which
the organisms that are carrying the gene reproduce themselves. In
the word "themselves" is the key to understanding the genetic appeal, as it were, of cloning. In sexual reproduction, a gene of one
of the parents has only a 50 percent chance of being reproduced;
with cloning, it is 100 percent. We might incautiously expect,
therefore, an evolved preference for cloning, similar to the evolved
preference of most people for their children (who have on average
50 percent of each parent's genes) over their nephews (who have
on average 25 percent of each uncle's or aunt's genes). Yet we do

not find a preference for cloning. The reason is that reproduction by cloning was not an available choice for human beings during the period in which the genetic makeup of the human race—the basis of our instinctual preferences and aversions—reached its present state. The likely reason that this choice did not evolve is that the reshuffling of the genes with every generation, which we get with sexual reproduction, provides protection against co-evolving parasites.[4] From the standpoint of inclusive fitness, the benefits apparently exceed the costs, for "natural" human cloning is limited to the rare case of identical twins.

The fact that a particular course of conduct might increase the frequency of one's genes doesn't mean that it will be undertaken. Otherwise the demand to be a donor to a sperm bank would be much greater than it is, for it is an extremely cheap way for a man to increase the frequency of his genes. Since there were no sperm banks in the period in which human beings evolved to their present state, a proclivity to donate to such banks has never evolved. Likewise there is no innate proclivity to clone oneself; but equally important, there is no innate aversion to cloning oneself, as there is to heights, which were, as cloning was not, a feature of our distant ancestors' environment.

The absence of an instinctual aversion is important because sexual desire is not the only evolved mechanism for stimulating reproduction. People love children, particularly their own; so adoption is rarely considered a perfect substitute for having natural children, even though the natural route will often be more costly for the mother. Parents enjoy noticing physical and mental resemblances between their children and themselves and thinking of their children as conferring upon themselves a kind of immortality. This narcissistic tendency, which we call evolved rather than acculturated because of its universality and its importance to reproductive fitness—people who don't have a strong preference for their own children are unlikely to produce many descendants—is likely to make some people, perhaps a great many people, desire perfect genetic copies of themselves. Very few people prefer to be the parents of the biological child of another person even if that child is greatly superior to what they themselves could produce, unless they have a deadly genetic defect. Adoption is a last resort.[5] Some people might therefore prefer to have a child that was entirely their own, rather than only half their own, from a genetic standpoint. This preference would be a logical extension of the well-documented tendency in animal species and primitive human communities to assist relatives in proportion to the fraction of shared genes.[6] That proportion reaches 100 percent for clones and identical twins.

In short, why share your genes if you don't have to? We are not likely to shudder at the thought of cloning ourselves, given the absence of an instinctual fear of cloning. As for danger from co-evolving parasites, modern medicine has largely banished that concern. Moreover, if it is a danger, it is one to the health of the human race as a whole rather than to that of an individual faced with choosing between sexual reproduction and cloning; the clone is unlikely to be more susceptible to infectious disease than his parent.

There are cultural as distinct from biological answers to the question, Why share your genes if you don't have to? but they would not convince everybody to follow the traditional route if cloning were cheap. One answer is through sexual reproduction you may produce someone even better than yourself, with the improvement compensating for the dilution of your genes in the next generation. This answer will appeal especially to people whose success in life exceeds what one would have predicted from knowing their genetic endowment. These people can "buy" the superior genes of a spouse with the financial resources or social prestige that is the fruit of their worldly success. Such a purchase is especially attractive from the standpoint of reproductive fitness when the purchaser has some genetic defect that will limit the reproductive capabilities of his clone.

Another answer to the question, Why dilute your genetic legacy?, is that it is a price of marriage—you will have to give your spouse a share of "your" children's genes. If this is an attractive trade, presumably because you put a high value on marriage or the particular marriage partner, it means that, as in the previous example, the dilution of your genes is compensated.

Both examples illustrate the important point that our genetic endowment does not completely determine our behavior. So from the fact that cloning would often be a way of maximizing the number of copies of our genes in the next generation it cannot be inferred that the demand for cloning will be great, even if the monetary cost is modest. Specifically, the demand for human cloning is likely to be concentrated in people who have "good" genes (by which we mean genes that make it more likely that a person will have good physical and mental health, high intelligence, or other prized talents, energy, and physical attractiveness, not necessarily genes that maximize reproductive fitness) and would not derive great benefits from marriage. These will sometimes, perhaps often, be the same people. Good genes as we have defined them are positively correlated with worldly success, that is, what makes them "good" in a society such as ours. The more successful a person is, the better able he will be either to marry on his own terms or to get along without being married at all. Some of these people will want to marry anyway; but others will not. Already we observe many people choosing not to marry. There would be more—and with a tilt not observed today toward the genetically and financially privileged—if human cloning were feasible and cheap. Cloning would thus be "anti-marriage," and, even if cheap, would benefit mostly rich men and women.

In stressing "normal narcissism" as a spur to cloning in cases where there is no problem of reproductive failure, we may have seemed to overlook a simpler point, that cloning provides a method of quality control or assurance. If we think of reproduction as the "purchase" of a child by its parents, the "product" cannot be observed before it is purchased or its qualities ascertained with any confidence. Cloning overcomes this uncertainty—or does it? The prospective parent may not be certain how many of his own qualities are due to his genes and how many to randomly favorable environmental factors that are unlikely to be duplicated in the upbringing of his clone child. He can reduce this uncertainty by mating with a person who has similar qualities, since the probability that the qualities of both persons are the product of luck rather than genes is less than the probability that the qualities of one of the two persons are.

From what we have said so far, it should be apparent that analyzing the demand for cloning and the social effects if the demand is allowed to be satisfied is difficult and involves many imponderables, even if the supply of cloning services is unproblematic. Intuition is not a reliable guide to estimating the consequences of cloning. Consider the most "obvious" of these consequences: an increase in the birth rate. By providing an alternative to sexual reproduction that some people might prefer, cloning would reduce the total costs of producing children. Yet the number of children might not increase. Cloning does not just reduce the cost of having a child, for example, to a person for whom sexual reproduction might be impossible or unappealing; it produces a different *kind* of child, namely, an identical twin of the parent. Someone who considered this kind superior to a child produced by sexual reproduction might decide to have fewer children, substituting perceived quality for quantity.[7] This would be especially likely if people generally prefer to have a child of their own sex, since cloning will produce that every time. Indeed, it seems plausible that people who cloned themselves would generally want to have just one child. The second child would be identical to the first, and a mixture of clones and sexually produced children might engender serious tensions. It is possible, therefore, that cloning would lead to a reduction rather than an increase in the birth rate.

We need a model to help us sort through these issues.

A Model of the Demand for Human Cloning

We begin by assuming that people seek to maximize their children's welfare,[8] viewed as an increasing function of the child's genetic endowment.[9] Imagine a society consisting of ten people, all adults. For simplicity, assume that everyone is of the same sex and can mate with anyone else and that each person has one child either by cloning or in the usual way; if the latter, the couple has exactly two children, to preserve the ratio of one adult to one child. Each person can be ranked from 1 to 10, with person 1 having the least desirable genetic endowment and person 10 the highest. A child is assumed to have the average genetic endowment of its parent(s); therefore, if a person clones himself, his child will have the same genetic endowment as he. Implicitly this assumes, but plausibly if we confine our attention to just a few generations, that the environment is not changing radically. If it is, the clone may be less well adapted than the sexually produced child, because the clone's missing parent may have genes better adapted to the new environment.

Table 1 reveals the payoffs under alternative reproductive regimes: a regime in which mating is the only option, a regime in which one may mate or be cloned, and a regime in which only people with a genetic endowment greater than 7 may clone themselves. This last option approximates a world in which the genetically best endowed are also the wealthiest and only the wealthiest people can afford to be cloned.

To understand the payoffs under mating, observe that 10 will marry 9, giving their child a genetic endowment of 9.5. This leaves 8 to marry 7, 6 to marry 5, . . . 2 to marry 1. Under cloning, 10 will clone himself because the payoff (10) exceeds the payoff from marrying 9 (9.5).[10] While 9 would rather marry 10 than clone himself, 10 is no longer available. But 9 would rather clone himself (and obtain a 9 child) than marry 8 and obtain for his (their) child

TABLE 1

Genetic Endowment of Offspring Under Alternative Reproductive Regimes

Parent	Reproductive Regime		
	Mating	*Cloning*	*Cloning >7*
1	1.5	1	0.0
2	1.5	2	2.5
3	3.5	3	2.5
4	3.5	4	4.5
5	5.5	5	4.5
6	5.5	6	6.5
7	7.5	7	6.5
8	7.5	8	8.0
9	9.5	9	9.0
10	9.5	10	10.0

an endowment of only 8.5. This process will continue all the way to 1, who must clone himself because there is no one left for him to mate with. When only the genetically best-endowed people can clone themselves (the last column in the table), all the less well endowed mate with each other unless, as in 1's case, no one is left for him to mate with.

The model suggests the possibility that the option to clone oneself could drive out sexual reproduction (except for the occasional contraception failure) and thus the mixing of genes over generations. The genetically best-endowed people in the model clone themselves because they do not want to mix their genes with people at the next level down. The people at the next level down do not want to mix their genes with the people below them, so they clone themselves as well. And this continues all the way to the least well endowed. Cloning would not completely displace sexual reproduction, however, unless it was possible to determine people's genetic endowments accurately, the preference for maximizing

one's children's genetic endowments overrode all other preferences, and cloning was not much more costly than sexual reproduction.[11]

Notice that the availability of free cloning would not necessarily help the genetically best endowed at the expense of the least endowed, as one might expect. It would make 10 better off, 9 worse
off, 8 better off, 7 worse off, and so on. The availability of free
cloning would make the least well endowed (types 1–5) worse off
as a group only if they would otherwise marry the best endowed
(types 6–10). They would not. The well endowed would generally
marry each other, in order to provide the best genetic endowment
for their children, and this would leave the least endowed to marry
each other.[12] Therefore, the availability of cloning would make
some well endowed better off and others worse off and some
poorly endowed better off and others worse off.

Even if cloning were expensive, so that only people with a genetic endowment (and, we are assuming, corresponding wealth)
greater than 7 could afford it, the best endowed might not be made
better off or the least endowed worse off. The availability of expensive cloning would make 9 worse off because it would allow 10
to remove himself from the marriage pool, eliminating 9's chance
of obtaining some of 10's genes for his offspring. It would also
make 7 worse off because 8 would clone himself. But 7 could no
longer marry 8 and so would have to marry 6, and this would make
6 better off than under either alternative regime.

A risk-neutral person, evaluating the regimes behind the veil of
ignorance, would thus be indifferent between no cloning and free
cloning but would prefer either regime to expensive cloning because the average payoff for the first two regimes is 5.5 and for the
third regime is 5.4. But this (slight) difference arises only because
we have assumed that an odd number of people can afford to clone
themselves in a society consisting of an even number of people, so
that person 1 cannot have any children. This is an artifact of the example.

A risk-averse person might prefer the no-cloning regime to the free-cloning regime and the free-cloning regime to the expensive-cloning regime. For notice that in the table the distribution of pay-offs widens as one moves from left to right. People might fear cloning because they do not like the idea that one could be born into a world in which one's children are certain to inherit one's bad genes, as opposed to one in which some mixing is likely. The advantage of mixing to the risk-averse person is that he gains more from avoiding the worst result (having the worst genes and passing them on unmixed) than he loses from not being able to achieve the best result (having the best genes and passing them on unmixed). But he would have to weigh this gain against the fact that the no-cloning regime forces him to bear the risk of infertility. If you're infertile, only through cloning can you transmit your genes to the next generation.

We can enrich the model by assuming that a child's welfare is an increasing function of his wealth (including the value of his education prior to adulthood and gifts and bequests from the parent afterward) as well as of his genetic endowment. We assume diminishing marginal utility both of genetic endowment and of wealth, so that an equal amount of each produces more welfare than do unequal amounts.

Imagine that society consists of 100 people, each of whom can be located within a 10-by-10 matrix, with genetic endowment on one axis and wealth on the other. Each person is assumed to have a unique genes-wealth pair, so that, for example, (1,1) denotes a poor person with bad genes and (10,10) a rich person with good genes. The average child produced in the usual way will have the average of his parents' genetic endowments, so that, for example, the mating of (10,10) and (2,4) will produce on average a (6,7), and for simplicity we'll now drop the qualification "on average" and assume that every child has the average of his parents' endowments. We define a person's welfare as the sum of the logarithms of each of his endowments to reflect the diminishing marginal utility of each. For example, welfare for $(9,1)$ is 0.95, whereas for $(5,5)$ it is 1.40.

Under these assumptions and in a regime of no cloning, the very highly endowed will marry each other and the least endowed will marry each other, but rich people with bad genes will marry poor people with good genes. $(10,10)$ does best by marrying $(9,10)$ or $(10,9)$, while $(1,9)$ does better by marrying $(9,1)$ than by marrying $(5,5)$, and $(5,5)$ does better by marrying $(6,4)$ than by marrying $(9,1)$. The match between $(10,10)$ and $(9,10)$ produces a child with endowments of $(9.5,10)$ and welfare, therefore, of 1.98. The match between $(1,9)$ and $(9,1)$ produces a child with endowments of $(5,5)$ and welfare of 1.40, and the match between $(5,5)$ and $(6,4)$ produces a child with endowments of $(5.5,4.5)$ for a welfare of 1.39, while a match between $(9,1)$ and $(5,5)$ produces a child $(7,3)$ with welfare of 1.32.

In a regime of free cloning, people with equal endowments would clone themselves (those on the high end by choice, those on the low end because no one would marry them), although people with unequal endowments would continue to marry each other. The results would not be much different in a regime of expensive cloning. Again, the high equals would clone themselves; the un-equals, even if wealthy, would marry; but this time the low equals would have to marry each other rather than clone. Sexual repro-duction would continue to be preferred by many people. People with good genes but little wealth would want to "trade" their genes for money in order to have the wherewithal to support and finan-cially endow their offspring, while wealthy people with poor genes would want to trade their money for genes. Both types of trade re-quire sexual reproduction. Yet on fairly ordinary assumptions about what people desire in their children, many people—all the equals—would clone themselves, and as a result the amount of genetic mixing would decline. And since people who had both great wealth and superb genes would no longer have to spread their wealth and genetic material in order to have children, cloning might foster the emergence of a genetic and financial elite.[13]

The Model Further Enriched

We can enrich the model further by asking, What if people could have as many children as they wanted? In a regime in which cloning is feasible and permitted, rich people with good genes who wanted to maximize the welfare of each child would have just one clone and no other children. The reason is that wealth, as distinct from genes, must be spread (though not necessarily evenly) among multiple children whether they are clones or the products of sexual reproduction. But if instead rich people wanted to maximize the chance that their genes would survive for many generations, the best strategy might be either to have multiple clones, because the extra genetic copies are cheap and only the wealth must be divided, or to have a few clones and a few ordinary children because a sexual partner with an overall poorer genetic endowment may still have some superior genes. A few people might wish to clone other people besides themselves or their children,[14] but we shall ignore this possibility. Nothing in our analysis depends on whether all people who want to clone want to clone themselves. A recent poll revealed that 6 percent of the respondents wanted to clone themselves, and apparently none wanted to clone anyone else.[15]

To the extent that cloning increased the demand for children by offering superior opportunities for maximizing one's influence on the gene pool, inequalities of wealth would decline because rich people would have more children among whom to divide their wealth. This possibility should moderate concern that cloning would increase disparities in wealth, genetic endowment, and overall welfare. But as we noted earlier, it is by no means certain that cloning would result in more children. Moreover, cooperation among clones, facilitated by genetic identity, which should reduce friction (as seems to be the case with identical twins), might enable the wealth of a family to grow faster than the number of members.

Let us consider differences between the sexes in the demand for clone. Cloning would benefit women more than men,[16] and so the demand for clone would presumably be greater among women. Cloning would allow them to have children later in life, enabling them to invest more in their market skills earlier.[17] Women while pregnant or taking care of the children would be less dependent on men for support. As a result, there would be fewer marriages; unmarried women would become wealthier relative to men; married women would have greater bargaining power in marriage. These effects would be multiplied when the woman had great wealth, good genes, or both. Wealth would allow her to raise the child alone, and good genes would make her less likely to find a man with equally good genes. Her benefits from marriage and sexual reproduction would be small.

Although the availability of cloning would benefit women with good genes and good market skills, it is less clear that it would benefit other women. On the one hand, cloning would hurt women by reducing men's demand for women's fertility. Because a man could have himself cloned and pay for help in raising the child, he could satisfy his desire to reproduce himself without enlisting the participation of a woman, save as the incubator of the clone fetus.[18] This would hurt women who lacked wealth or good genes, since such a woman would not be able to compensate the man for sharing his children's genes with her by forgoing cloning. On the other hand, cloning would benefit women by reducing their dependence on men's fertility. Because a woman could have herself cloned and pay for household help, she could satisfy her desire to reproduce herself without enlisting the aid of a man. This option would be available to poor women if welfare paid for the costs of raising clone children. Moreover, cloning by wealthy women would increase the demand for womb rental by poor women, since there would be little reason for a wealthy woman to carry her clone fetus herself, especially with no husband to help out.

Gillian Hadfield argues that women and girls (maybe at the urging of their parents) invest in skills that are complements to the skills ordinarily possessed by men because women with complementary skills are more desirable marriage partners than women with redundant skills.[19] As evidence, she points out that in all societies men and women specialize in different kinds of work, but that with some exceptions for work requiring great strength there is little cross-cultural consistency in the kind of work that men and women do. The availability of cheap cloning would reduce the importance for women of having complementary market skills. Girls would no longer be as likely to invest in complementary education; they would invest in whatever education would maximize their lifetime earnings independently of a husband's career. The result would be an even more rapid entry of women into areas of the workforce traditionally dominated by men than we are observing today.

Human cloning might thus portend an accelerating breakdown in the traditional roles of men and women and facilitate the emergence of a class of wealthy and powerful women—both disturbing prospects to men and women who hold traditional views of sex roles.

To summarize the discussion to this point, cloning would benefit mainly wealthy women with good genes and to a lesser extent wealthy men with good genes. One would therefore expect, if human cloning were feasible and permitted, a growing concentration of wealth and highly desired heritable characteristics at the top end of the distribution of these goods and fewer marriages there. Although the rest of the population distribution would be made relatively worse off as a group, many people within that part of the distribution would be made better off, including people with incurable infertility.

But the model is still too abstract. Wealth, genetic endowment, risk aversion, fertility, and sexual difference are not the only important variables bearing on the demand for and consequences of cloning. Here is a mundane but frightening point: The demand for cloning would be disproportionately concentrated in people whose narcissism exceeded normal bounds, and, more generally, in people who today are prevented from (or rather impeded in) reproducing by being unmarriageable, usually because of severe personality disorders. Normal people want to mate with other normal people, not with people who are psychotic; and psychotics themselves probably do not want to mate with other psychotics, and often do not want to mate or associate with anyone, since difficulty of establishing personal relationships is a symptom of a disordered personality.[20] Extreme narcissists in particular would probably not want to marry anyone, save on terms intolerable to any self-respecting person[21]—especially another narcissist! Other types of men and women who today have difficulty finding mates include mentally retarded people, people with serious physical disabilities, convicted felons, homosexuals, pedophiles, and sociopaths. Men despairing of or rejecting marriage (or simply wanting to have more children than is feasible through sexual reproduction in a society that outlaws polygamy) who wanted to clone themselves would still have to rent a womb, and that would create some constraint, even though the necessity of finding a mate would be eliminated. Women who cloned themselves would be self-sufficient; they would have merely to bear the cost of pregnancy. Concern about clones carrying defective genes and raised by disordered persons might engender pressure for governmental screening of people who wanted to clone themselves, thus raising the spectre of eugenic regulation. This would be an example of how technology, by eliminating a social or biological barrier to an activity, can increase the optimal scope of government.

Persons with extraordinary talents having a large genetic component, such as champion athletes and world-class musicians, might be tempted to clone themselves. If so, then over time the economic rents obtained by persons who have scarce and highly valued genetic endowments would decline—an income-equalizing effect of cloning. An esoteric but important class of potential demanders for cloning are dictators, who might believe that problems of succession would be lessened if a clone were waiting in the wings. Imagine if when Stalin died, a fifty-year-old Stalin clone had been Stalin's designated successor; imagine if today Fidel Castro had a fifty-year-old clone.

An important variable in the demand for human cloning is the desire of most people to marry. As we noted earlier, they are unlikely to be able to do so if they are "gene selfish." In addition, most people do not seek to produce a child who is *merely* financially and genetically well endowed, but one who is happy, and most people believe that happier children have two parents. What is more, because of economies of scale and specialization within the household, it is less than twice as expensive for a couple to raise two children than for a single parent to raise one. Against these points it can be noted that the desire to marry is in part a function of the desire for children. The more the desire for children can be satisfied by alternative arrangements, the less demand there will be for marriage. And cloning can be reconciled with marriage and dual parenting in the following way: The married couple can decide that rather than producing two children sexually they will each clone. This is not a perfect solution for them. Because a person is more closely related to his or her clone than to his or her sexually reproduced children, and *a fortiori* than to his or her spouse's clone, to whom indeed he or she is not related at all, each spouse may have difficulty thinking of himself or herself as a parent of both children; so dual cloning may not produce dual parenting.

We can put this differently. The man who "sells" his wife a genetic half-interest in "his" children gets in return more than someone who will take a share (maybe the lion's share) in the rearing of the children. He gets a child rearer who has a superior *motivation* to do a good job precisely because of the genetic bond. Altruism is a substitute for market incentives, and the man can take advantage of this substitute by giving his wife a genetic stake in the children.[22] So marriage and sexual reproduction would remain for many, probably for most, persons a superior alternative to cloning, even if cloning were not only lawful but also very cheap.

Many people, moreover, want to have children that differ from them in important (genetic) respects. They wish to improve their stock, which they cannot do by cloning themselves, or to hedge against the risk that their own genes are not as good as they think. Even if they are preoccupied with, or driven subliminally by their genes to maximize, their inclusive fitness, they can do this as well by having two children, each of whom shares half their genes, as by having two children, one of whom shares all their genes and the other (the spouse's clone) none.

And it is impossible to know whether people would find cloning an attractive option until we know what a clone would be like. A clone might seem disappointingly different from his parent, or eerily similar; in either case, people might prefer sexual reproduction. And we have not taken into account possible social responses to cloning. If cloning led to an extraordinarily unequal distribution of wealth, society might respond by imposing highly progressive taxes. It might even place an excise tax on cloning. Then adverse effects on wealth distribution would not be compelling arguments against the availability of cloning—unless the costs of social measures to reduce the distributive effects of cloning were great, which they might be.

In a discussion of an imagined but no longer unforeseeable reproductive technology that would allow a husband and wife to choose which of their genes to give to their child, Thomas Schelling points out that people might compete over characteristics.[23] They might, for example, choose taller children in the hope of giving them competitive advantages. But their hope would be dashed because all children would become taller, assuming many other parents also had a preference for tall children. Because cloning gives people less power over the genetic characteristics of their children, the danger of such a zero-sum competition is less. But the general point still holds. To the extent that genetic endowment is a positional good, competition over it does not produce social gains; in contrast, competition in the market produces social gains because market goods are, for the most part, nonpositional, at least if envy is ignored.

Schelling's point raises the general question of the effect of cloning on the clones themselves. Earlier we assumed that parents want to maximize their children's welfare. This is an unrealistic assumption. Rational parents want to maximize their own welfare, and thus their children's only to the extent that the children's welfare enters into the parents' utility function. So we cannot assume, at least when people have a choice between cloning and sexual reproduction, that their children's welfare will be maximized by the choice made.

Setting to one side biological uncertainties that we assume will eventually be dispelled,[24] the clone will be a perfectly normal human being, as normal as an identical twin. But the vertical relation of genetic identity has different implications from the horizontal relation. Take the case in which a married couple decides to have two clones, one of each spouse, rather than producing children sexually. If the clones then clone themselves, the original husband and wife will have the same genetic relationship to their grandchildren as to their children, while their children will have no genetic relationship to each other and also their grandchildren no genetic relationship to each other. If the (unrelated) children marry each other and coproduce a child, the original husband will have a closer relationship with his grandchild than with the cloned child he has through his wife. Or suppose a husband and a wife coproduce a child and then clone the child while he is still an infant. Is the clone the child's sibling or the child's child? Is he the father's child or the father's grandchild? If the clone grows up and clones himself, the original husband and wife will have the same genetic relationship to their grandchild as they have to their child. In these examples, cloning might run up against the deep-seated incest taboo, though this is speculative.

In the Very Long Term

We consider, finally, some highly speculative long-term effects of human cloning. One is that it might reduce the genetic diversity of the human race by facilitating eugenic breeding. Imagine: Parents coproduce a child, who at the age of three manifests signs of great precocity. They clone this child rather than coproduce another child. Or parents have two children and clone the better-looking or more intelligent one. Fertile parents who share a common genetic defect or infertile parents who have genetic defects may choose to clone superior relatives or, indeed, to purchase the right to clone other people who have desirable genes, although that tendency will be retarded by the preference for own over adopted children. Because cloning involves a smaller genetic gamble than does a combination of sperm and egg of even highly desirable strangers, cloning would be preferred to artificial insemination or surrogate motherhood by those attracted to the idea of selective breeding. To the extent that selection was in favor of a few widely desired features, and against the widely undesired, human genetic diversity would decrease, with obvious risks to human adaptability to unforeseeable changes in the environment that might make currently undesirable traits more valuable and currently desirable traits less valuable.

In cultures in which boys are valued more than girls, parents might decide to clone the father or, having coproduced a son, to clone that son rather than risk having daughters. Over time, sex ratios could change dramatically.[25] This process may be self-correcting in the long run, because girls will become more valued offspring as the ratio of males to females rises. Even so, the inevitable lags in the self-correcting process might cause grave social dislocations and incite demands for intrusive government regulation of reproductive decisions,[26] for which we now have ample precedent in East Asia. Asked to correct an undesirable sex ratio, government would have to choose some legal instrument. Maybe it would tax the cloning of men but not the cloning of women, or tax cloning but not coproduction. Since wealthier people would have more clones than poorer people, a tendency accelerated by the tax, wealthier people would have relatively more boys. And once it became acceptable for the government to influence cloning, could interest groups resist using the government to encourage the cloning of some people (geniuses?) but not others (the genetically defective)? We would then be in the much-feared world of eugenic regulation.

Earlier we showed that cloning might have a tendency to crowd out sexual reproduction. The more people clone, the fewer people are available for sexual reproduction; the pool of potential reproductive partners shrinks, and it becomes more difficult to produce a superior child by reproduction than by cloning. A different path of crowding out is opened if we consider the possible long-term consequences of even the relatively benign public policy with which we began this essay: permitting only infertile couples to clone themselves. Currently, mutant genes that interfere with sexual reproduction cannot be propagated; infertile people do not have offspring. When cloning becomes available, the genes that enable sexual reproduction will lose their survival advantage over genes that interfere with sexual reproduction. Imagine a society consisting of 50 men and 50 women. Assume that in every generation 2 percent of the people (for simplicity, one man and one woman) are infertile because of a condition that is heritable. Assume everyone marries and the average couple has 2.04 children (for example, 48 couples have two children and one couple has 4 children). In a world without cloning, the genes of the infertile people will not be reproduced, and every generation will replace itself by producing

100 children. Now suppose that cloning becomes available. Each generation's infertile couples will have on average 2.04 (cloned) children, who will be infertile (for we are discussing cases in which infertility is caused by an inherited defect). So while the first generation will consist of 98 fertile people and 2 infertile people, the second generation will consist of 98 fertile people and 4 infertile people, and the third generation of 98 fertile people and 6 infertile people. In five generations clones would constitute almost 10 percent of the population; eventually they would be dominant.[27]

It is not clear whether sexual reproduction would eventually disappear. On the one hand, random mutations would interfere with sexual reproduction but would not interfere with the cloning of infertile people—a crucial asymmetry. And as the percentage of fertile people fell, the costs of matching would rise because the population of potential mates would be small. On the other hand, the mixing of genes that results from sexual reproduction may enhance survival, even under the environmentally gentler conditions brought about by modern medicine.

As long as clones must be incubated in human wombs (which may not be for long, for artificial wombs are being developed), infertile men and women would generally (depending on the nature of the woman's infertility) have to pay fertile women[28] to bear their clones. If as a result these women did not reproduce themselves, infertility would spread even more rapidly than in our numerical example. It is true that if infertile individuals married fertile individuals and the couple decided to have only the fertile partner cloned, the genes for infertility would not be reproduced. But it is more likely that the infertile partner would demand that at least one child be his or her clone; the preference for costly and painful reproductive technologies over adoption attests to the importance that people attach to genetic reproduction (what we earlier discussed as normal narcissism). So each infertile person might make clones of himself or herself without matching up with anyone else, or might marry another infertile person and share the burden of raising two clones.

People can clone themselves faster than they can produce children through sexual reproduction, which imposes a delay of more than nine months between children. Because of the strong incentives of modern women to delay childbearing until late in life, this might give cloning a great advantage over sexual reproduction. Perhaps great enough to increase the birth rate of the infertile over the fertile, assuming cloning were practiced only by the former. To take an extreme example, if infertile couples clone themselves once (that is, produce two clones) in the time that a fertile couple takes to have one child, and if time between births is the only constraint on reproduction, then starting in our world of 100 people of whom 2 are infertile, infertile clones would outnumber fertile people in about five generations.

Our estimates of the possible effects of cloning on fertility are, of course, highly sensitive to the percentage of persons having *heritable* infertility. The percentage is not known. It is undoubtedly only a small fraction of all persons who are infertile, because genes that cause infertility are maladaptive and hence highly likely to be selected out in the course of evolution. The number of persons with fertility problems, heritable or nonheritable, is unknown, because only people who are trying and failing to have a child (and not all of them) seek medical attention for such problems. Infertility, moreover, is often a function of the couple, each member of which might be fertile with another sexual partner. It has, however, been estimated that at least 20 percent of fertility problems are male and that 10 to 20 percent of these are genetic.[26] Assuming a like percentage of genetic female infertility problems (for which, however, we have not been able to find any substantiation), 10 to 20 percent of all fertility problems are genetic. The higher figure may be consistent with the estimate we used earlier, that 2 percent of couples have heritable fertility problems. An estimated 7.1 percent of married couples have fertility problems,[30] and this is clearly an underestimate, not only because noncomplainers are not counted but also because people who know themselves to be infertile are less likely to marry. On the other hand, fertility problems often merely delay rather than prevent conception and birth, and often are treatable; so 2 percent may be too high after all. But since random genetic mutations can cause a fertile person to become infertile, while it is extremely unlikely that a random mutation would cause an already infertile person to become fertile, it may not be crucial whether 2 percent or 0.2 percent or even 0.002 percent of the population is afflicted with mutations that impair infertility. Infertility will spread like a virus, merely at different rates, and eventually drive off fertility.

The spread of infertility through cloning might be even more rapid if, as realism requires, "reproductive failure" were defined broadly enough to encompass the situation of a homosexual couple, for whom cloning might be an attractive alternative to adoption, artificial insemination (if it is a lesbian couple), or surrogate motherhood (if it is a male homosexual couple). Assuming that all or most homosexual orientation is genetic,[31] the fraction of homosexual genes in the gene pool would be increased if cloning resulted in a disproportionate increase in reproduction by homosexuals, who might be thought of as "functionally" infertile to the extent that they do not reproduce sexually. But this depends on the transmission path of the homosexual gene. If the gene predisposing to male homosexuality is through the female line,[32] then male homosexuals will not transmit the gene to their clones.

Is the spread of infertility throughout the population something to be feared, when, by assumption, people are able to reproduce using cloning technologies? As noted earlier, the evolutionary advantage of sexual over asexual reproduction is that the mixing of genes protects future generations against coevolving parasites. If everyone cloned himself, future generations would have the same genetic diversity as the current generation; so parasites that evolved the capacity to crack the immunological defenses of members of the current generation would pose a threat to the members of future generations who were their clones. And if some people cloned themselves more than others, future generations would have less genetic diversity than the current generation. Genetic diversity, like vaccination, is a barrier to the spread of parasites. Like the person who refuses to be vaccinated, the person who clones himself or herself does not internalize all the costs of his or her behavior. Unless medical technology evolves as quickly as parasites do, over time the human race could find itself increasingly vulnerable to disease.

Conclusion

Our exploration of the likely demand for human cloning has been strictly that—exploratory. The demand is impossible to estimate; it depends on too many variables of uncertain strength. But the analysis does provide a rational basis for the widespread disquiet that the prospect of human cloning has aroused. Some of that disquiet has religious or emotional foundations that our analysis does not touch; some of it reflects an unreasoning fear of change. But consider: The most sympathetic demanders for human cloning, the infertile, may, over time, if allowed to clone, drive out sexual reproduction. The least sympathetic demanders, extreme narcissists and other psychotics and misfits, will be among the most enthusiastic for cloning, and their cloning too will feed on itself to the extent that the disorder that makes them unmarriageable is hereditary.[33] The point is not that cloning frees each sex from dependence on the other, though it does (women more clearly than men, however, since a womb is still necessary), but that it eliminates the barrier to reproduction that is created by the need to find another person willing to mate with you. That barrier is a screen against reproduction by people with serious maladjustments.

Cloning may also aggravate inequalities in genetic endowment and in wealth, undermine the already imperiled institution of marriage, alter the sex ratio, and create irresistible pressures for eugenic regulation. This is on the one hand. On the other hand, some of the frightening effects of cloning may be offsetting: If, as we speculate, cloning will increase the wealth and power of women, the demand for daughters may rise, canceling out a preference for sons that cloning might enable parents to indulge. And some of the effects are so long run that technological advances of the very kind that have given us cloning may eliminate them: Long before the population becomes dominated by infertile and narcissistic clones, infertility and extreme narcissism may be as passé as smallpox. In other words, fertility technology and psychiatric medicine may advance as rapidly and as far as cloning technology. Perhaps, then, despite the concerns discussed in this paper, only the very cautious will want to prohibit human cloning.

Endnotes

1. See I. Wilmut *et al.,* "Viable Offspring Derived from Fetal and Adult Mammalian Cells," 385 *Nature* 810 (1997), and for a popular treatment Ruth Macklin, "Human Cloning? Don't Just Say No," *U.S. News & World Report,* March 10, 1997, p. 64. The technique involves replacing the nucleus of an ovum with the nucleus of a cell of the animal to be cloned. The ovum is then implanted in a womb, where it grows into a baby in the normal way. When we speak of "cloning" in this essay we mean the method by which Dolly was created, that is, by the cloning of adult nonreproductive tissue from a single animal or human being. The possibility of cloning an adult human being gives new meaning to the term "single parent."

2. See "Whatever Next?" *Economist,* March 1, 1997, p. 79.

3. See, for example, Leon R. Kass, "The Wisdom of Repugnance: Why We Should Ban the Cloning of Humans," *New Republic,* June 1, 1997, p. 17. For an earlier treatment, when cloning was foreseen but not yet within reach, see Paul Ramsey, "Shall We Clone a Man?" in *Fabricated Man: The Ethics of Genetic Control*, ch. 2 (New Haven: Yale University Press, 1970).

4. See William D. Hamilton, "Sex versus Non-Sex versus Parasite," 35 *Oikos* 282 (1980); Robert Trivers, *Social Evolution* 322–330 (Menlo Park: CA: Benjamin/Cummings Publishing, 1985).

5. See, for example, Linda S. Williams, "Adoption Actions and Attitudes of Couples Seeking In Vitro Fertilization: An Exploratory Study," 13 *Journal of Family Issues* 99 (1992), and studies cited there.

6. See, for example, Matt Ridley and Richard Dawkins, "The Natural Selection of Altruism," in *Altruism and Helping Behavior: Social, Personality, and Developmental Perspectives* 19 (J. Philippe Rushton and Richard M. Sorrentino, eds.; Hillsdale, NJ: L. Erlbaum Associates, 1981); Trivers, note 4 above, chs. 3, 6, 15.

7. On the quantity-quality tradeoff in children, see Gary S. Becker, *A Treatise on the Family* 145–154 (enlarged ed.; Cambridge, MA: Harvard University Press, 1991).

8. The assumption is obviously unrealistic, implying as it does that, beyond some age, parents would transfer all their wealth to their children and starve. We relax the assumption later.

9. Another unrealistic, temporary assumption: It abstracts from other constituents of welfare, such as financial resources, not due solely to one's genetic makeup.

10. Notice the assumption that 10 *knows* he's a 10. If there is uncertainty about one's genetic fitness, one may decide to hedge one's bets by mating with someone of similar qualities, as we noted earlier.

11. Another threat posed by cloning to the future of sexual reproduction is considered later in the essay.

12. On the tendency to positive assortative mating, see Becker, note 7 above, at 112–118. With the breakdown in the United States of traditional cultural barriers to marriage between persons otherwise alike (such as barriers against crossing religious or ethnic lines), assortative mating is increasingly likely to take a genetic form. Yet even when such barriers are insurmountable, assortative mating along genetic lines takes place behind the barriers, that is, within the segmented groups.

13. The Hardy-Weinberg Equilibrium, in which gene frequencies remain unchanged from generation to generation, assumes random mating. See, for example, Eli C. Minkoff, *Evolutionary Biology* 142–144 (Reading, MA: Addison Wesley, 1983); Mark Ridley, *Evolution* 87–90 (Boston: Blackwell Scientific, 1993). That is not a realistic assumption for human beings.

14. Humbert Humbert, for example, might have wanted to clone Lolita.

15. "Clone the Clowns," *Economist,* March 1, 1997, p. 80.

16. This is the general effect of technological improvements in reproduction. See Richard A. Posner, "Separating Reproduction from Sex," in *Sex and Reason,* ch. 15 (Cambridge, MA: Harvard University Press, 1992).

17. Although they can get the same result by freezing their eggs and hiring a surro-

gate mother to incubate them, there is still the problem of obtaining sperm to fertilize them. The woman may not want to go to the bother of finding a man with good genes to be the father. She can avoid the bother by going to a sperm bank, but then she is taking a genetic gamble.

18. A current in ancient Greek thought represented by Aeschylus and Aristotle. They believed that *all* children were the father's clone—that the woman's role in reproduction was limited to incubation.

19. Gillian K. Hadfield, "A Coordination Model of the Sexual Division of Labor" (unpublished, University of Toronto Law School, 1996).

20. See, for example, American Psychiatric Association, *Diagnostic and Statistical Manual of Mental Disorders* 638–642 (4th ed.; Washington, DC: American Psychiatric Association, 1994).

21. See, for example, Arnold Rothstein, *The Narcissistic Pursuit of Perfection* 87 (New York: International Universities Press, 1980).

22. Important evidence for this is the enormously increased risk of child abuse by stepparents compared to parents. See Martin Daly and Margo Wilson, *Homicide* 83–93 (New York: A. de Gruyter, 1988).

23. Thomas C. Schelling, *Micromotives and Macrobehavior,* ch. 6 (New York: Norton, 1978).

24. These are lucidly described in the *Economist* article, note 2 above. For example, it is uncertain whether a mammal cloned from nonreproductive tissue would have a normal lifespan; the clone's biological age might be the sum of its and its parent's chronological ages.

25. See Schelling, note 23 above, at 197–203.

26. *Id.* at 202–203.

27. *Cf.* Michael Bliss, *The Discovery of Insulin* 245 (Chicago: University of Chicago Press, 1982): "Because insulin enabled diabetics to live and propagate, and because the disease had a strong hereditary component, the effect of the discovery of insulin was to cause a steady increase in the number of diabetics."

28. Fertile in the sense of being able to carry a fetus to term; they might be infertile in the sense of being unable to produce an egg. The discussion in text assumes that all women are either fertile or infertile in both senses.

29. D. M. de Kretser, "Male Infertility," 349 *Lancet* 787 (1997).

30. Joyce C. Abma *et al.,* "Fertility, Family Planning, and Women's Health: New Data from the 1995 National Survey of Family Growth," 23 *Vital Health Statistics,* no. 19, p. 7 (Centers for Disease Control and Prevention, May 1997).

31. For a summary of the evidence, which however is largely limited to male homosexuality, see Richard A. Posner, "The Economic Approach to Homosexuality," in *Sex, Preference, and Family: Essays on Law and Nature* 173, 186, 191 n. 26 (David M. Estlund and Martha C. Nussbaum, eds.; New York: Oxford, 1997).

32. See Dean H. Hamer *et al.*, "A Linkage between DNA Markers on the X Chromosome and Male Sexual Orientation," 261 *Science* 321 (1993).

33. One can also envisage a demand for clones on the part of people who want a source of "spare parts" for organ transplants and other medical needs.

The authors thank Héctor Acevedo-Polanco, Emlyn Eisenach, Sorin Feiner, Gertrud Fremling, Dan Kahan, Leo Katz, William Landes, Martha Nussbaum, Charlene Posner, Reed Shuldiner, Robert Trivers, and participants in the University of Chicago's Workshop on Rational Models in the Social Sciences for many helpful comments on a previous draft of this essay, and Acevedo-Polanco, Feiner, and Brady Mickelsen for helpful research assistance.

A Rush to Caution: Cloning Human Beings

Richard A. Epstein

What Me Worry?

ew announcements have provoked more rapid public fascination
and academic dismay than the news story in the *Observer* on February 23, 1997: Ian Wilmut and his colleagues at the Roslin Institute
had successfully cloned Dolly, a sheep from a cell drawn from her
(as it were) mother's mammary glands. That successful cloning
clearly represented a major and somewhat unexpected technical
breakthrough, and since that day further advances have followed
rapidly. On July 25, 1997, the *New York Times* reported that a team
led by Dr. Wilmut and Dr. Keith Campbell had been able to create a lamb that had a human gene in every cell of its body.[1] That result was quickly topped by news that a Wisconsin biotech company
had been able to clone three identical calves from fetal calf tissue,
using processes that it claimed were more efficient and reliable
than those used by the Wilmut group.[2]

My own position, which I shall develop here, is that these new
developments call for no immediate legal response: Watchful waiting is far preferable to hasty or ill-conceived legislation whose
unanticipated consequences are likely to do more harm than good.
First, do no harm, is as good a principle now as it has ever been.
But inaction leaves political actors on the sidelines, where they be-

long but not where they would like to be. So it was all too predictable that the first reports on cloning whipped into action a
powerful coalition of the bioethical and legal professions. Various
nations across the world, and several states in the United States
took the first steps toward a ban on cloning human beings, and,
somewhat more diffidently, research that could lead to the cloning
of human beings. President Clinton responded to the whiff of crisis first by imposing a temporary ban on federal funding of cloning
research, and then by asking President Harold Shapiro of Princeton University to head a distinguished review team of the National
Bioethics Advisory Commission (NBAC) to chart the legal and political responses to cloning after taking into account the ethical
and religious concerns raised about the practice. The report was
duly prepared within 90 days and issued these two recommendations:

· A continuation of the current moratorium on the use of federal
  funding in support of any attempt to create a child by somatic
  cell nuclear transfer [i.e., cloning].
· An immediate request to all firms, clinicians, investigators, and
  professional societies in the private and nonfederally funded sectors to comply voluntarily with the intent of the federal moratorium. Professional and scientific societies should make clear
  that any attempt to create a child by somatic cell nuclear transfer and implantation into a woman's body would at this time be
  an irresponsible, unethical, and unprofessional act.[3]

In addition, the NBAC recommended that any legislation prohibiting cloning should contain a sunset clause of between three and
five years—an eternity in the field of biotechnology. The NBAC
also urged that the regulations in question "be carefully written so
as not to interfere with other important areas of scientific research."[4] It then enigmatically suggested that "if the legislative ban

is ever lifted," the twin protections of independent review and informed consent be used to protect the human subjects who participate in preliminary research trials on cloning. From the tone of the report the "ever" suggests that "never" is the preferred position of most of the ethicists and theologians who have anxiously considered the problem.

Why adopt this position in preference to watchful waiting? The usual justifications examine the various ramifications of cloning. One set of objections is quickly proving to be transitory, namely, that the practice is too dangerous to be conducted on human beings today. The second set of objections—those favoring "never"— promise a more permanent ban: a wide set of religious and ethical misgivings that promise to become more, rather than less, insistent as the techniques of cloning are improved.

But "why?" is a question that should be asked in a second way: Why impose the ban *now?* The NBAC's report spends a good deal of time on the merits of human cloning, but it spends far less time thinking through the logic of research bans, whether induced by government decree or moral persuasion. That issue is worth a few comments here. My basic position is that our rush toward caution is not warranted by the factual information, practical doubts, religious convictions, or moral intuitions invoked to sustain it. I plan to get at this issue by taking a slight detour. I shall examine what I consider to be the basic conditions for banning any practice, and thereafter apply these standards to human cloning. I do not wish at present to commit myself to a position that cloning should be forever legal. Still less do I wish to commit myself to a view that some vision of untrammeled reproductive rights places human cloning on some preferred constitutional plane that defeats any and all government efforts to regulate or forbid the practice. Rather my point is merely one of practical philosophy: The presumption of liberty of action counsels waiting, not rushing, to impose any legal prohibitions on human cloning research.

The Basic Framework

It is always dangerous business to set out a writer's basic framework in a few short paragraphs, but I do so to orient the reader.[5] I think that the rights and wrongs of human action are not determined by one grand insight, but come from a set of successful approximations that move us cautiously but surely toward some social ideal. The first step on this journey is that individuals are ordinarily entitled to do what they want. Now that someone (the actor) benefits, the presumption of free action can be overridden only by showing negative consequences to other individuals. Even here, however, a lot depends on the nature of the supposed harm: killing, raping, and maiming are one thing; individual harm from open competition is quite another. The former should galvanize the state into action. The latter should provoke a collective yawn, because over the long run competitive harms are inseparably linked to the competitive benefits that on net advance human welfare. So until we run into monopoly power, the case for liberty of action remains intact.

The analysis thus far presupposes that we can accurately track the consequences of individual actions or practices, so that the major social task is to divide legal from illegal conduct. Our task becomes more complex when we don't know the actual consequences of these actions or practices. At this point we—and here it is the collective "we" to which I refer—are forced to separate forbidden from permitted actions. But we are unable to make the division on the intrinsic quality of certain types of acts: What matters is how things turn out in individual cases. X is about to drive an automobile. Surely we should stop him if we know he is going to crash into a house or run over a pedestrian. Yet surely we should bid him fond farewell if we know he will arrive safely at his destination. But what should we do when we are unclear whether his journey will end in disaster or success?

At this point, it is best to examine a range of factors to identify the occasions and ways to apply collective force. One factor, obviously, is the expected cost of the feared harm, which, roughly speaking, is defined as the probability that harm will occur to some individual, multiplied by its anticipated severity. The greater the likelihood of harm, and the greater its expected severity, the stronger the case for banning the action in question. But we also have to consider a second type of error, that of banning activity that might prove harmless or even beneficial: What is the expected probability of a successful outcome, multiplied by the social gain that comes from its occurrence? Even armchair estimates of the relevant variables remind us that life has few certainties. For a start, at least we should resist any rule that bans all activities whose expected costs of harm are greater than zero, wholly without regard to the offsetting benefits: Nothing would ever get done, and we would also starve, unless we could somehow ban starvation as well. Both benefits and costs count in reaching some sensible *final* position on banning.

From that simple observation many nettlesome complications follow. One factor worth looking at is the distribution of gains and losses between persons. Self-inflicted harms usually present little reason for legal intervention: Nature's feedback mechanism is strong enough to stop most self-destructive conduct. It is harm to others, not harm writ-large, that provokes social concern. Now the soundness of any ban depends in large measure on what will happen if no ban is imposed. If the individual actor is on the hook for money damages, the case for banning those activities is weakened: The risk of financial loss will have some of the cautionary effects of self-inflicted injury. To be sure, it takes a bothersome lawsuit to make the damage payment stick, but against institutional defendants with substantial assets, that threat is real. So the basic rule is where damage actions look as though they are effective, then the case for a ban is correspondingly weakened.

More importantly, we should always be alert to choices in the size and scope of the proposed ban. We could ban everyone from driving a car; we could ban only those who are under sixteen years old; or we could also ban people over sixteen years old who have not passed a test, or who have committed serious traffic violations. Where a more restrictive ban is possible in the future, we should be cautious about imposing the ban today. Better to let the useful activity go forward, and then select the spots in which to impose the ban. In addition, we can mix partial bans, licenses, and damage actions to get the right levels of driving, hopefully by the right people. These principles, moreover, generalize to cover not only driving cars, but also clearing swamps, building apartment houses, and spraying disinfectants. Risk is endemic to all activities; even so—or better, precisely so—we should only shut down those activities that hold a high risk of substantial peril or harm. The strategy will backfire on occasion, and some actions will take place that we will regret in the fullness of time. But the simple fact that we have automobile accidents offers no justification for banning cars, and the occurrence of high-profile oil spills gives no reason for banning oil tankers from the high seas, although it might justify routing them out of environmentally sensitive areas. Deciding what bans to impose and who should implement them is part science and part art. Some judgment calls are close, but throughout all the initial presumption should always be set in favor of liberty of action: Be careful and go slow with the ban.

Banning or Waiting?

How does this framework carry over to the distinct issue of cloning? To answer that question we have to identify and weigh the relevant variables on both sides of the ledger: the harms that might occur, and the probability of their occurrence, versus the benefits that could be obtained and the probability of their occurrence.

The difficulties hit us at the first step. The implicit assumption of the earlier model was that the relevant harms fell into certain definable classes: bodily injury or death to persons; damage to private property, or even to public lands and waters. But cloning presents none of the customary forms of harm that have been the target of legal action. Indeed the term "identifiable harm" seems sadly out of place for the diffuse set of perils at stake. Indeed I find it something of a mystery why cloning is said to give rise to some natural and widespread human revulsion which, if we follow Leon Kass, should serve as us our moral lead.[6] I confess some curiosity about how it would all work out and a little edginess about the worst-case scenarios. But we should keep this in perspective. Cloning does not belong on a list with assault, incest, illegitimacy, debauchery, deformity, depravity, or neglect. What we have is the creation of one new person whose genetic component is identical to that of another human being, alive or dead. What, we may ask, is the harm in all that?

The simple, hard-line answer sticks with the narrowest conception of harm and concludes nothing at all. That approach makes our decisions easy, perhaps too easy: Cloning should never be banned. Once we are certain that the practice in question involves no harm (as defined) to other persons, then the expected losses are zero: Offense, or even revulsion, toward cloning, and the satisfaction, or even curiosity, that others might attach to it are both ignored. So long as someone wants to undertake the activity, its expected gains are positive. This uncluttered libertarian view thus ends the matter: If there is no threat of future damages, then there is no case for a present ban.

The most cursory review of the NBAC report indicates that its perplexities begin exactly where this simple libertarian approach ends. The somewhat depressive report contains a menu of real and fancied harms that cannot be ignored. Start with the question of psychological harms. We have the newly cloned infant who is now told that he is genetically identical with some other human being— perhaps his "father" or "mother" (whose quotation marks need no explanation), or perhaps some public personage of great intellectual, athletic, or artistic merit. What would we expect his reaction to be to his subordinate and derivative place? Here the implicit assumption of the anti-cloning forces is that the distressed newcomer would have been in some sense better off if he had never been born: This situation reminds me of the unhappy experience I had in the Yale Law School moot court competition in arguing with a straight face that a child with severe damages from German measles contracted during pregnancy would have been "better off" never having been born at all.[7] But rather than try to plumb these philosophical depths, I will take the criticism as asking, Why invest in a cloned individual when we could find someone else who did not have to carry this psychological baggage around for all the days of his life?

So now the issue about the burdens heaped on the new person is just empirical. How heavy is the load? I confess to substantial ignorance on the question. Today we know of lots of people who are clones of each other: the entire class of identical twins, some of whom were reared together and others who were reared apart. And here the evidence is decidedly mixed. Some twins might lack the space that comes from being one of a kind. On the other hand, many twins report that they enjoy a special kinship that comes from knowing that another person shares their moral, aesthetic, and emotional presuppositions. It helps to get rid of the loneliness in facing the world. My own casual research led me to a story in the August 1997 issue of *Marie Claire,* which features two male identical twins who met and married two female identical twins at a twins convention in Twinsberg, Ohio, and then set up house together as a foursome.[8] Apparently no psychological baggage here. Clearly some twins rejoice in their twinned state. But recall that the question is not whether we would prefer to be a singleton or twin. It is whether we should ban cloning right now.

Perhaps the bonding between natural identical twins is not decisive on the issue of psychological harms. These twins are of the same age: One does not have to live in the shadow of those who have gone before, and that element could well bring about a certain level of gloom and frustration. Who could hope to live up to an older identical clone, or keep a sense of optimism and independence if that person should suddenly suffer a deadly disease or commit some heinous act? it might be hard to make plans for one's own future, when we have uncommon insight as to what it will be. The NBAC quotes the philosopher Hans Jonas who warns against cloning as violative of the right to ignorance, a right which I for one would gladly waive about tomorrow's closing stock prices. The philosopher Joel Feinberg is mentioned as speaking about the right to an open future, and the philosopher Martha Nussbaum is taken to defend the importance of separateness of persons.[9]

These sound like very lofty concerns. But how are they likely to play out in practice? Everyone concedes that the genetic identity of two individuals predisposes them to act in parallel ways. Will the clone of Michael Jordan regard life as a failure if he does not reach the basketball heights of, as it were, his genesake? But before becoming excessively gloomy on the subject, note that other outcomes are possible. The children of successful parents often have greater confidence in their own ability to succeed. Perhaps the clone would say, "Look, with my genes I have a real advantage over those other children of distinguished parents who face the strong tendency of regression toward the mean." The future holds lots of uncertainties even for people who know their genetic complement. Some people might like to have a leg up on others who are struggling to figure out their own identities. Perhaps the clone would have, dare one suggest it, the psychological edge.

Either result is, in my view, possible once we recall the determinants of personal identity and success. The new individuals are genetic twins of their single ancestor, but different people. As the NBAC stresses, the thought that we could clone "a team of Michael Jordans, a physics department of Albert Einsteins, or an opera chorus [no soloists?] of Pavarottis, is simply false."[10] We do not have to indulge in speculation about multiple cloning to see why no copy of the parent is likely to be just that person. The basic explanation is almost too simple: Human behavior is a function of genetic endowment interacting with social and environmental forces. Those who reach the top at anything need a fair bit of luck to do it. A twisted ankle at age four, a disagreeable science teacher in third grade, a sore throat in winter could throw the best-endowed clone permanently off the fast track, or result in the development of some previously ignored talent. Major differences in uterine environment (which separates the clone from the identical twin) are enough to account for big differences in personality and intelligence;[11] differences in diet and care, in fad and fashion, in occupation and education should also yield perceptible differences between "ancestor" and clone and preclude exact duplication of personality. Given the confluence of change and determination, similarity and difference, who can confidently predict the psychological response, positive or negative, of the cloned person to his second-chair status, with or without counseling?

There is of course no direct evidence, but perhaps a parallel provides some food for thought. Cloning is but one of many high-tech methods of reproduction. Other methods also torment bioethics today.[12] We also have artificial insemination (father or donor), in vitro fertilization, and surrogate parenthood. And we know how to manipulate the contents of individual cells early in the reproductive process. Are we aware of any major social and psychological damage to the offspring of these practices? Here we actually do have some evidence, and it speaks to the adaptability of human beings, not to their psychological fixations.

Moving further afield, surrogacy involves the artificial insemination (not strictly necessary for the practice to survive) of a woman who is paid to nurture an unborn to birth and then turn it over to the biological father and his own spouse. Sounds pretty weird. The literature contains many abstract philosophical denunciations of how this dangerous form of commodification—how I hate this use of the word!—will lead to the impoverishment of these young lives.[13] But the evidence does not seem to support that conclusion. Professor Lori Andrews did her own informal survey of surrogacy and found that the women hired as surrogates were able to explain the position quite comfortably to their own children and to get on with their lives.[14] Obviously, the practice is not for all. Yet it is by allowing persons to sort themselves out by contract, and to hold them to the contracts that they make, that the institution seems to have flourished.[15] The stability of the contractual arrangement is not dispositive of the position of the children born to these arrangements. But the critics of surrogacy have not presented any evidence that the children born of these surrogate arrangements show any special psychological problems. Given the resources of their parents, and their determination to go through with this procedure, my guess is that these children will usually be reared in circumstances that children naturally conceived in less fortunate circumstances would envy. The calls for ban in surrogacy have met with partial success. Yet it is far less clear that they have been justified by any showing of psychological harm to any of the relevant parties.

That lesson should carry over to the cloning case. My speculations could be morally obtuse; or the finely honed anxieties of the philosophical and theological elites could be unrepresentative of how most people will respond to the practice of cloning if it comes their way. The variation in human responses is likely to be wide, and I have some confidence at least that the people who will make themselves candidates for cloning (either by using their cells or carrying the child) are more likely to be well equipped for the emotional ride than those who recoil from the practice and thus keep their distance.

So we have lots of uncertainty about the matter, and that is precisely why it is so problematic to justify a ban on harm-prevention grounds. The usual legal standard requires clear evidence of serious harm, at the least. Here we do not come close to meeting it.

If we were to wait a bit, and see what happens, then we could form our judgments with greater knowledge as to how the practice did impact the lives of the newly cloned and their immediate families. At that point, we might learn things that support the worse fears of the opponents of cloning, and if so, then we can use hard evidence, not abstract fears, to decide what form of ban (whole or partial, conditional or absolute) to impose on the practice. It may turn out that the only clones who have trouble are those of a person who assumes the role of parent, or those of sports figures, or whatever. If so, the narrower ban could be tailored to the abuse. Knowledge, not ignorance, could guide our deliberations. No one says that this course of action is risk-free. It could well be that matters turn out well at the beginning and the harms begin only years later when many clones have been born. But again the point cuts both ways. To treat these fears as decisive is to say that we can never have cloning unless we have certain knowledge of the future. That cost could be high as well. Our task is to minimize error, not to cling to the status quo.

The same pattern of argument carries over to other concerns. Cloning surely makes it difficult to maintain the usual social roles of parent, sibling, grandparent, nephew, and niece. And we could easily join Leon Kass and others who brood about the level of disconnectedness that comes by awkwardly shoehorning this new person into a set of relationships created by nature and instituted by countless cultures over thousands of years. Again, it may seem presumptuous to respond so cavalierly to these epic themes, but still it is worth suggesting in a small voice that most human beings have the internal flexibility to adapt to this novel situation. We know that adoptions take children out of their natural families and place them in very different settings; interracial adoptions are still more dramatic, and yet many of these adoptions are the source of the most profound satisfaction to parents and child alike.[16] Are resourceful parents and a responsive legal system wholly unequal to the challenge?

Furthermore, we do not have to quibble over legal issues that can be resolved right now. If the question is whether the child is person or object: Person is the emphatic answer, with rights equal, as both child and citizen, to those of any other child however conceived and wherever raised. It is here that separateness matters. We can make it clear by statute and by agreement, if not by birth, that the woman who gives birth to a cloned child must agree to raise it as her own, for better or for worse. We can demand that her husband take on the obligations of its father. For single women, we might be forced to do without a father, as we now do when these women adopt or give birth to their own children. The conscious assumption of these obligations is no small matter, and leads people to reconsider their actions before they undertake them.

No one can guarantee that fixing the legal position will ensure an ideally affective and emotional upbringing. But no abstract concerns should allow the best to become the enemy of the good. Lots of children born within ordinary families come into horrific social settings. Forcing couples and single women to think hard about the legal position should encourage desirable social practices that assimilate the cloned child to traditional patterns of childrearing; my guess is that the child will worry less about cosmic identity and more about sports or fashion. So long as that outcome is not improbable, we should hesitate before banning the practice on the strength of unsubstantiated fears that it will miscarry. So, on this score too the ban is premature: Lesser precautions, many easily implemented, can reduce some of the risk, and allow the experiment to be played out a bit longer.

The proponents of the ban find it easy to think of other forms of harm. Human beings who evolve by sexual reproduction have a diversity that could not be achieved by asexual reproduction, which is why all higher forms of life are built around these sexual differences. That diversity has strong benefits for the overall survival of the human race: To put all of our eggs (quite literally) in one basket violates this fundamental norm of diversification. In good times the undiversified species will flourish, but in bad times the species could be wiped out, so that good times will never come again. To follow the apt language of the financial analyst: Diversification of the portfolio gives us the best hedge against the uncertainty of the future.

Yet I can see no reason to fault cloning on grounds of incomplete diversification. Cloning is not the same practice as inbreeding: The guess is that any clones will come from successful individuals, not failures, and we should not expect to see with each passing generation the rise of recessive traits that will cripple the species. Nor is it clear how much, if at all, cloning could reduce diversity. The cloned individual will in all probability mate with other persons of his or her own age. Cloning thus opens up new possibilities for unique genetic combination as it forecloses others. I doubt that anyone could find any appreciable difference in the composition of the gene pool even if cloning becomes a widespread practice.

Finally, here lies the rub. What makes anyone think that cloning will become widespread even if legalized? My guess is that it will make some dent on human reproductive practices in a few desperate cases. But in the grand scheme of things, the impact is likely to be slight. It is too quickly forgotten that allowing cloning does not require anyone to clone. The opponents of cloning may, and should, raise their theological and ethical arguments to all who will listen, and inveigh against the practice with all the passion they can summon. (I am happy to report that they will be able to persuade me not to partake in the grand experiment in any role.) In all likelihood, given the sense of professional opinion, they will have some success with others: Even Wilmut and his fellow cloners are more interested in using cloning for genetic experimentation in agriculture and livestock than in making human beings. We know that cloning humans presents certain risks, even though it already seems possible to do better than 1 successful cloning in 277 attempts, the success rate with Dolly. No matter how great the advances, the process will likely remain delicate so that, just a guess, the odds of successful completion will be well below 50 percent. If the law forces individuals to bear the costs of their own vanity, then price itself will deter many from striking out in this novel direction. We can expect to see few takers given the other available alternatives. Why therefore worry about the effects on biodiversity of a practice that is not likely to affect even 1 person in a million? For those of us who want to worry, start with illegitimacy, foster care, abuse, neglect, and starvation, where some intelligent human intervention might improve what has become a baleful situation.

In response it could be asked, If the likelihood of cloning is so low then why object to the ban at all? Few will be inconvenienced and the greater certainty in the legal and social environment will work to the long-term advantage of us all. But the balance of convenience runs clearly in the opposite direction. No ban costs little to enforce: We do not have to draft regulations that "carefully distinguish" between the research practices that are banned and those that can continue. We do not have to worry about the international cooperation that is necessary to make sure that the practice does not take hold in Scotland, or for that matter, Singapore or Qatar. And we do not have to worry about the possibility that the ban will, in spite of our best intentions, prevent the kind of practices that give very little reason for concern. Already I have been told that biotech companies that have no corporate interest in getting into the human cloning business are concerned by the overbreadth of the potential legislation, which could interfere with their own more focused efforts to develop new treatments for various genetic disorders. And they should be. The bills introduced in the House and Senate this past March had the virtue of being short; they had the greater vice of being wholly uninformative. One bill provides: "It shall be unlawful for any person to use a human somatic cell for the process of producing a human clone."[17] Another bill is still broader in scope: "No Federal funds may be used for research with respect to the cloning of a human individual."[18] It offers no definition of what counts as "research," even if it does offer a definition of what counts as cloning.[19] The pithy nature of the bill will require supplementation through regulation, but why assume that bureaucrats and administrators will issue regulations that take a narrow view of their power?

Here is one example. Suppose that some great benefit from cloning comes not with Michael Jordan, but with work with infertile couples. Suppose that they are able to produce fertilized eggs, and that these for some reason have to be reproduced. Suppose further that cloning might help in this process. If so, then we have a ban that surely overshoots its mark. This cloning is not of a present human being. We do not have to worry about individual autonomy, unrooted individuals, preordained futures, psychological deprivation, or biodiversity. If the cloning works we have a new unique child with two traditional biological parents. The placement of the child is where we want it, even if it reached that end by a devious path. Who knows whether this path will prove viable, but introducing the ban may well knock it out.

So wait, and learn. If it should come to pass that some horror stories occur, then we can think of more focused responses, just as we do in so many other areas of life. Let ten thousand impressionable members of the same Girl Scout troop decide that they want to carry the precise likeness of their favorite (and willing) heartthrob; then even I would think that a narrow ban on (to use the pejorative term) "trafficking" might be appropriate. Let some famous rock star auction off cells ripe for cloning, and perhaps I could be persuaded to go along with the tidal wave of sentiment that would call for stopping the practice (or at least allowed for a cooling-off period), perhaps even before the first sale took place. But that is just the point. If we can wait, then we can focus, and if we can focus then we can eliminate the chaff without knocking out the wheat. So my advice is to continue the debate while we allow the research to go on full steam ahead. The new information should allow us to make better decisions, but only if we take a more level-headed and less angst-driven approach to the problem than seems to have dominated the gloomy NBAC. The end of the human race is not at hand; so for the time being, relax.

Endnotes

1. Gina Kolata, "Lab Yields Lamb with Human Gene," *New York Times,* July 25, 1997, at A12.

2. Ronald Kotulak, "Move over Dolly: U.S. Firm Clones in New Way," *Chicago Tribune,* August 8, 1997, at 1.

3. NBAC, Report ("Cloning Human Beings," 1997) at 108–109. (Hereafter called NBAC.)

4. *Ibid.* at 109.

5. For the longer account, see Richard A. Epstein, *Simple Rules for a Complex World* (Cambridge, MA: Harvard University Press, 1995).

6. Leon Kass, "The Wisdom of Repugnance," *New Republic,* June 2, 1997, at 17.

7. The case was *Gleitman v. Cosgrove,* 49 N.J. 22, 227 A.2d 689 (N.J. 1967).

8. See Lucy Broadbent, "Twin Set," *Marie Claire,* August 1997, at 40.

9. NBAC at 67–68. Nussbaum is clearly misunderstood. Her position is only that all individuals deserve dignity and respect of their own. She takes no position on whether there is anything objectionable in having two persons with identical genetic profiles.

10. NBAC at 33.

11. See Sharon Begley, "Wombs with a View," *Newsweek,* August 11, 1997, at 61.

12. Gina Kolata, "Scientists Face New Ethical Quandaries in Baby-Making," *New York Times,* August 19, 1997, at B7.

13. See, e.g., Elizabeth Anderson, *Value in Ethics and Economics* (Cambridge, MA: Harvard University Press, 1993).

14. Lori Andrews, "Beyond Doctrinal Boundaries: A Legal Framework for Surrogate Motherhood," 81 *Va. L. Rev.* 2343 (1995).

15. Richard A. Epstein, "Surrogacy: The Case for Full Contractual Enforcement," 81 *Va. L. Rev.* 2305 (1995).

16. See Janny Scott, "Orphan Girls of China at Home in New York," *New York Times,* August 19, 1997, at A1, on the adoption of Chinese girls by middle-aged New York professional couples.

17. H.R. 923 (introduced March 5, 1997).

18. S. 368 (introduced February 27, 1997).

1 9. "(b) *Definition*—For purposes of this section, the term 'cloning' means the replication of a human individual by the taking of a cell with genetic material and the cultivation of the cell through the egg, embryo, fetal, and newborn stages into a new human individual." *Ibid.*

I should like to thank L. Rex Sears and Michael Maimin of the Law School Class of 1999 for their valuable assistance in preparing this article.

On Order

Barbara Katz Rothman

Cloning is about control. It's about introducing predictability and order into the wildly unpredictable crapshoot that is life. If normal procreation is the roll of a hundred thousand dice, a random dip in the gene pool, cloning is a carefully placed order. And that's where it gets interesting: It is *order* both in the sense of predictability and control, and in the sense of the market, an order placed, a human being on order.

In a perfect world, we could think about the value of the first form of order, the value of predictability and control in procreation, without thinking about the second form of order, the power of the market. In our world, the two are hopelessly, endlessly entangled. I personally am not convinced that predictability and control are really achievable in human procreation, nor am I convinced that they would be good things to achieve. But those points are at least open for debate, for discussion. I am completely convinced that market forces are an evil in human procreation.

That leaves me in the funny kind of place I often am with the new technologies of procreation: Thank goodness they don't work terribly well. The only thing that could make them worse would be if they got better.

Predictability has its limits in procreation. Think for a moment about identical twins, "nature's own clone" as it were. Identical is a funny word. It means they're the same. And so they are, strikingly so, which is why people noticed that they are a particular kind of twin. Identical twins arise from the same fertilized egg. One zygote becomes two people. Or even more rarely, three. At the moment of zygotic zero, when the egg and the sperm join, the twins are not only identical, they are one and the same. Very soon after that they split and go their separate ways.

If the splitting occurs within the first three days, they each develop their own placentas. If splitting occurs a bit later, between the fourth and the eighth day, they, each within its own amniotic sac, share a placenta. If the splitting is later still, they will share a single sac too. And later than that, the twins themselves may be joined, "Siamese" twins, two people in a more-or-less shared body.[1]

How identical are identical twins? Right there, in the very same woman, nestled in the very same womb, they go their separate ways, experiencing life differently. Genetic changes take place as the eternal splitting and building of cells occur. Mutations can occur differently in each. One could be born a boy, with the full X, Y chromosomes; the other, missing the Y chromosome in many of its cells, a girl. A girl, with one X and no second sex chromosome, a condition known as Turner's syndrome, can be the "identical" twin of her brother, grown of the same fertilized egg.[2]

And even without dramatic mutations occurring, twins are not the same baby twice. At birth they do not weigh the same: Their rates of growth can be very different. Placed here or there, in the same womb in the same woman in the same environment, they're not having the same experience. *Here* isn't *there,* and nothing is ever the same.

They don't even have the same "genetic" diseases. Type I diabetes, for example, has been traced to a gene located on the small arm of the sixth chromosome. One variation of that gene, one allele, "causes" autoimmune diabetes. In one twin. In one identical twin the gene causes diabetes and in the other the same gene does not cause diabetes. If one twin has type I diabetes, the chances of an identical twin having it are only 30 percent.[3] More than two-thirds of the time the identical twin will not have this "genetic" disease. Some genes are more powerful, in which case the odds of a shared disease or trait go up; some are less powerful, and the odds go down. Diseases, traits, characteristics—two people, two individuals from one fertilized egg. You can't predict which will be which, how it will play itself out over time.

You can't get away from the idea of chance. Start with the same egg and the same sperm forming the same zygote in the same woman, and one twin gets sick and asks "Why me?" Clone the same cell and each person produced will be different from the source of the cell, and different from each other. No matter how much we want to control, to predict, to answer that question of "Why?" the answer sometimes really is just *because,* luck, chance.

That is *not* what we used to think science was about. Physicists have had to come to terms with that uncertainty, raising it to a principle. Classic physics thought that you could know it all, "that our cosmos is governed by mathematically precise laws at all scales, from inside of an atom to the totality of the universe."[4] And then chance raised its head, and "the positions and movement of the invisible atoms that make up all objects were now lost in a probabilistic blur. While the universe remained determinate, the revised mathematical constructs of physics—more accurately reflecting the way atoms behave—could no longer promise completely predictable events and objective, universal knowledge."[5] You can't, it seems, know it all.

Predictability and control are forever slipping out of our hands, crumbling as we touch them. When you're working with small herds of sheep, say, cloned to produce some expensive, exotic protein in their milk for medicinal purposes, a certain percentage of error is to be expected, accounted for. It is accounted for in two senses: It is part of the "account" or narrative of what might happen. (Shit happens!) And it is accounted for in the ledger books, an anticipated expense. Most of the errors that can be expected will not matter: Only a few will affect the single and sole purpose for which the sheep were cloned: the production of that protein in their milk.

With people, the accounting gets a lot more complicated, in both senses. Errors are not to be written off, and our expectations are rarely so narrowly confined.

We—those of us who have been around the track a few times—know how this latest "advance" in human procreative technology is going to be brought to us, marketed to us. It is going to be the solution to some heartrending problem. There's going to be a very good reason to do it the first time: some irreplaceable bone marrow donation, or some man who is the very, very, very last of his family, preferably made so by some cataclysmic political tragedy like the Holocaust. The guy will have absolutely, totally, no sperm whatsoever. Or maybe it will be . . . But I don't want to do this, don't want to give anybody any ideas.

For that first time, any success will probably be success enough. Later though, if we begin to make cloning routine, offer it as a service at the growing number of fertility clinics, the expectations will be more specific and at the same time more generalized. People will want to predict and control the kinds of children they are creating, or why bother using the technology?

And the "kinds of children" they might create are understood to be products of the nucleus of the cell from which they are created. Genetics—not just as a science or a set of technologies, but genetics as an ideology, as the way we are more and more thinking about life—is the descendent of classical patriarchal thinking. In a patriarchy, a father-based kinship system, men are the source of life; women the vessels. Life on earth is the product of the many and varied seeds of the earth; human life is the product of the seeds of man. The earth itself, in this thinking, is only the place, the location, where seeds are planted. The very words we use for earth— dirt, soil—indicate the disdain in which location is held relative to the glorious, powerful life-giving seed. In our more enlightened, modified patriarchy, women are also recognized as having seeds, but the primacy of the seed remains. Seeds count.

The idea of cloning suggests that the essence, the true essential quality, the very thing that makes a being itself—what once might have been called the soul—lies in the nucleus of its cells. Everything else is reduced to environment, background, ground, earth, dirt.

To hear the talk of the geneticists, one would think that the contents of that cell nucleus, the hundred thousand or so segments of DNA that constitute the "genes," are themselves the miracle of life. All life-forms begin in DNA; all life then is DNA; and maybe all life is, is DNA. They—the geneticists who have pushed to translate the full length of DNA into the letters GCAT so that it can be "read" and "mapped"—sound religious, awestruck, overwhelmed by the power and the majesty of the DNA: It's the Bible, the Holy Grail, the Book of Man. Those are their very words. Our fate, James Watson tells us, lies not in our stars but in our genes.

On a more mundane, work-a-day level, the DNA is called a code, an encyclopedia, an instruction kit, a program, and most often, a blueprint. Well, sure. I started out as a little cell with a nucleus of DNA and here I am. Those must have been plans in there.

But it took about fifty years to get where I am now, and I think that's worth thinking about. When you have the code or the plans for a person, you don't have the person. If you take my very DNA and clone it, make copies, and let those copies grow into people, it will take fifty years for them to get where I am now. And a lot happens in those fifty years.

If I want to build a cabin, and I have a blueprint, I could gather a bunch of people and machines and do it in a few days, or I could work with hand tools by myself on weekends for the next ten years. A blueprint doesn't have time or growth built into it and so a blueprint isn't really a good analogy for DNA.

A recipe might make more sense if we want analogies. Baking bread, for instance, combines making something with growth, the growth of the yeast that gives bread its rise. I've baked a lot of bread in my time, and I've learned that the same recipe under different circumstances results in different breads. Use a flour from a wheat grown in one part of the country and you have a different mineral composition and a slightly different taste than if you use flour from a wheat grown somewhere else. Bake on a humid day and you get a heavier bread from the same recipe than on a dry day. Bake on a hot day and the bread rises faster and has bigger air holes. Bake the same bread every day for a week, and no two loaves will be exactly the same: The web, that distinctive pattern of holes, will vary from loaf to loaf. Bake it in different pans or in different ovens, and you'll have differently textured crusts.

I've also made a couple of babies in my life. And while I understand about the DNA and the plans and the blueprints and all that, I also remember the rituals of avoidance. I read the ingredients on every food package before I used it, hesitating to put things I couldn't pronounce into the growing body of my baby. I avoided coffee, and cat litter dust, and, by the time of the second pregnancy when the rules had changed, alcohol. I read books about fetal development until I decided it was totally incapacitating: How could I possibly walk down the subway steps, cross a street and breathe exhaust fumes, even get out of bed, on arm-bud development day? All that blueprint following wasn't happening somewhere else. It was happening in my belly, day after day after day after day.

It's that element of time that seems so strangely absent in the discussion of cloning and DNA, in the thinking of genetic determinists. One of the Nobel laureates in genetics likes to hold up a compact disk at his lectures and say "This is you." As if the genes that were there at the moment of zygotic zero when I began fifty years ago, and the me that stands here now, were one and the same.

In the talk of the geneticists, time seems to exist only in terms of evolution, of progress, not in its daily, processual, experienced way, the time in which I bake a bread, grow a baby.

A blueprint is always located in a place and in a time. The same DNA, the same genes, will make a tall person or a short one, depending on the nutrition available during growth. One could have all the symptoms of copper deficiency because of an error in the gene to use copper—or because of a copper-deficient environment. Someone noticed the brittle wool of sheep grazing on copper-deficient land and identified the inability to metabolize copper as the cause of the genetic disease that has among its symptoms, brittle hair. The same DNA will produce sheep with lush wool in one place, and short unusable wool in another. The same DNA, dipped out of the same gene pool, will produce the short people of my grandparents' generation or the tall ones of my children's. Monday's bread is lighter than Tuesday's; Tuesday's crust is crispier than Wednesday's.

And what does the market introduce into this? What do we get when we take these variable processes of growth and time and turn them over to the forces of the market? Wonderbread: a nearly perfectly predictable bread.[6]

I cannot afford to do what my great-great-grandmother did out of necessity, bake my own bread, any more than I could afford her uncontrolled fertility. I can't afford the time. Home-baked bread, like eight children, is a luxury well beyond the likes of me. I almost always buy my bread, baking only for a treat, for a holiday, for the occasional pleasure of it. I can afford to buy a more customized bread than Wonderbread: at the bakery, at the farmer's market. I am buying the services of the baker, considerably more industrialized than what I could do at home, somewhat less industrialized than the factory supplying my supermarket. And sometimes, when I'm rushed or broke, I do grab those plastic wrapped loaves off the shelf.

Could people ever be mass-produced like supermarket loaves?

I don't actually think so, but the image frightens. The factories toss all the errors—the loaves that come out misshapen, that failed to rise evenly, that are burnt a bit or are underdone. It happens. They take that into account. It is part of their quality-control program.

The current technologies of procreation introduced first quantity control and more recently, and increasingly, quality control.[7] Prenatal screening and testing with the expectation of selective abortion are a form of quality control, avoiding the production of children we claim we can no longer afford to raise. The choice of quantity control, the choices offered to us by (relatively) safe and (relatively) effective contraception, eventually lost us the choice of not controlling the quantity of our children: Who could really afford the eight children my great-grandmother bore? And so it is with quality control: The introduction of that choice may ultimately cost us the choice not to control the quality, the choice of taking our chances in life's great, glorious, and terrifying roll of the hundred thousand dice.

Cloning may well eventually be offered to us as a way of avoiding the tragedy of prenatal diagnosis and selective abortion, the grief of deliberately ending a wanted pregnancy because of the kind of child it would produce.[8] Cloning may eventually—and eventually isn't as long a time as it used to be—be offered to us as a way of inserting predictability and control earlier in the process. Placing order in procreation: Placing our orders.

I don't think they'll be able to get that level of control, because I don't think life lives in that nucleus. If the DNA is a bible, it's capable of being read in a lot of different ways. My friend Eileen Moran says, "Think of the DNA as a musical score, notes on a page, but capable of nuanced interpretations." Is it possible that this static thing, this string of ACGTs, these notes on the page, *is* life? Or is life the way in which those notes are played?

And what is a bread? Is it the recipe? The ingredients? The process? Or isn't it all of those, the end product of a complex interaction played out over time?

And what are our children? Not seeds grown up, DNA transcribed, but also the product of a complex interaction played out over time.

So I'm not *that* worried about cloning. I don't think it can ever work terribly well. Not well enough to bank on.

References

1. Pritchard, Jack A., and Paul C. MacDonald. *Williams Obstetrics.* 16th ed. New York: Appleton Century Crofts, 1980, p. 640.

2. *Ibid.,* p. 644.

3. Burr, Chandler. *A Separate Creation: The Search for the Biological Origins of Sexual Orientation.* New York: Hyperion, 1996, p. 220.

4. Pollack, Robert. *Signs of Life: The Language and Meanings of DNA.* Boston: Houghton Mifflin, 1994, p. 150.

5. *Ibid.,* p. 150.

6. My appreciation to Maren Lockwood Carden for showing me where the bread-baking analogy was headed.

7. My appreciation to Roslyn Weinman for this insight and this wording.

8. For a fuller discussion of the difficulties and consequences of prenatal screening—for which cloning may be offered as the solution—see Barbara Katz Rothman, *The Tentative Pregnancy: How Amniocentesis Changes the Experience of Motherhood.* New York: W. W. Norton, 1993.

Recommendations of the Commission

National Bioethics Advisory Commission

With the announcement that an apparently quite normal sheep had been born in Scotland as a result of somatic cell nuclear transfer cloning came the realization that, as a society, we must yet again collectively decide whether and how to use what appeared to be a dramatic new technological power. The promise and the peril of this scientific advance was noted immediately around the world, but the prospects of creating human beings through this technique mainly elicited widespread resistance and/or concern. Despite this reaction, the scientific significance of the accomplishment, in terms of improved understanding of cell development and cell differentiation, should not be lost. The challenge to public policy is to support the myriad beneficial applications of this new technology, while simultaneously guarding against its more questionable uses.

Much of the negative reaction to the potential application of such cloning in humans can be attributed to fears about harms to the children who may result, particularly psychological harms associated with a possibly diminished sense of individuality and personal autonomy. Others express concern about a degradation in the quality of parenting and family life. And virtually all people agree that the current risks of physical harm to children associated with somatic cell nuclear transplantation cloning justify a prohibition at this time on such experimentation.

In addition to concerns about specific harms to children, people have frequently expressed fears that a widespread practice of somatic cell nuclear transfer cloning would undermine important social values by opening the door to a form of eugenics or by tempting some to manipulate others as if they were objects instead of persons. Arrayed against these concerns are other important social values, such as protecting personal choice, particularly in matters pertaining to procreation and childrearing, maintaining privacy and the freedom of scientific inquiry, and encouraging the possible development of new biomedical breakthroughs.

As somatic cell nuclear transfer cloning could represent a means of human reproduction for some people, limitations on that choice must be made only when the societal benefits of prohibition clearly outweigh the value of maintaining the private nature of such highly personal decisions. Especially in light of some arguably compelling cases for attempting to clone a human being using somatic cell nuclear transfer, the ethics of policy making must strike a balance between the values society wishes to reflect and issues of privacy and the freedom of individual choice.

To arrive at its recommendations concerning the use of somatic cell nuclear transfer techniques, the National Bioethics Advisory Commission (NBAC) also examined long-standing religious traditions that often influence and guide citizens' responses to new technologies. Religious positions on human cloning are pluralistic in their premises, modes of argument, and conclusions. Nevertheless, several major themes are prominent in Jewish, Roman Catholic, Protestant, and Islamic positions, including responsible human dominion over nature, human dignity and destiny, procreation, and family life. Some religious thinkers argue that the use of somatic cell nuclear transfer cloning to create a child would be intrinsically immoral and thus could never be morally justified; they usually propose a ban on such human cloning. Other religious thinkers contend that human cloning to create a child could be morally justified under some circumstances but hold that it should be strictly regulated in order to prevent abuses.

The public policies recommended with respect to the creation of a child using somatic cell nuclear transfer reflect the Commission's best judgments about both the ethics of attempting such an experiment and our view of traditions regarding limitations on individual actions in the name of the common good. At present, the use of this technique to create a child would be a premature experiment that exposes the developing child to unacceptable risks. This in itself is sufficient to justify a prohibition on cloning human beings at this time, even if such efforts were to be characterized as the exercise of a fundamental right to attempt to procreate. More speculative psychological harms to the child, and effects on the moral, religious, and cultural values of society may be enough to justify continued prohibitions in the future, but more time is needed for discussion and evaluation of these concerns.

Beyond the issue of the safety of the procedure, however, the NBAC found that concerns relating to the potential psychological harms to children and effects on the moral, religious, and cultural values of society merited further reflection and deliberation. Whether upon such further deliberation our nation will conclude that the use of cloning techniques to create children should be allowed or permanently banned is, for the moment, an open question. Time is an ally in this regard, allowing for the accrual of further data from animal experimentation, enabling an assessment of the prospective safety and efficacy of the procedure in humans, as well as granting a period of fuller national debate on ethical and social concerns. The Commission therefore concluded that there should be imposed a period of time in which no attempt is made to create a child using somatic cell nuclear transfer.

Within this overall framework the Commission came to the following conclusions and recommendations:

I. The Commission concludes that at this time it is morally unacceptable for anyone in the public or private sector, whether in a research or clinical setting, to attempt to create a child using somatic cell nuclear transfer cloning. We have reached a consensus on this point because current scientific information indicates that this technique is not safe to use in humans at this time. Indeed, we believe it would violate important ethical obligations were clinicians or researchers to attempt to create a child using these particular technologies, which are likely to involve unacceptable risks to the fetus and/or potential child. Moreover, in addition to safety concerns, many other serious ethical concerns have been identified, which require much more widespread and careful public deliberation before this technology may be used.

The Commission, therefore, recommends the following for immediate action:

· A continuation of the current moratorium on the use of federal funding in support of any attempt to create a child by somatic cell nuclear transfer.

· An immediate request to all firms, clinicians, investigators, and professional societies in the private and nonfederally funded sectors to comply voluntarily with the intent of the federal moratorium. Professional and scientific societies should make clear that any attempt to create a child by somatic cell nuclear transfer and implantation into a woman's body would at this time be an irresponsible, unethical, and unprofessional act.

II. The Commission further recommends that:

· Federal legislation should be enacted to prohibit anyone from attempting, whether in a research or clinical setting, to create a child through somatic cell nuclear transfer cloning. It is critical, however, that such legislation include a sunset clause to ensure that Congress will review the issue after a specified time period (three to five years) in order to decide whether the prohibition continues to be needed. If state legislation is enacted, it should also contain such a sunset provision. Any such legislation or associated regulation also ought to require that at some point prior to the expiration of the sunset period, an appropriate oversight body will evaluate and report on the current status of somatic cell nuclear transfer technology and on the ethical and social issues that its potential use to create human beings would raise in light of public understandings at that time.

III. The Commission also concludes that:
- Any regulatory or legislative actions undertaken to effect the foregoing prohibition on creating a child by somatic cell nuclear transfer should be carefully written so as not to interfere with other important areas of scientific research. In particular, no new regulations are required regarding the cloning of human DNA sequences and cell lines, since neither activity raises the scientific and ethical issues that arise from the attempt to create children through somatic cell nuclear transfer, and these fields of research have already provided important scientific and biomedical advances. Likewise, research on cloning animals by somatic cell nuclear transfer does not raise the issues implicated in attempting to use this technique for human cloning, and its continuation should only be subject to existing regulations regarding the humane use of animals and review by institution-based animal protection committees.
- If a legislative ban is not enacted, or if a legislative ban is ever lifted, clinical use of somatic cell nuclear transfer techniques to create a child should be preceded by research trials that are governed by the twin protections of independent review and informed consent, consistent with existing norms of human subjects protection.
- The United States Government should cooperate with other nations and international organizations to enforce any common aspects of their respective policies on the cloning of human beings.

IV. The Commission also concludes that different ethical and religious perspectives and traditions are divided on many of the important moral issues that surround any attempt to create a child using somatic cell nuclear transfer techniques. Therefore, we recommend that:

· The federal government, and all interested and concerned parties, encourage widespread and continuing deliberation on these issues in order to further our understanding of the ethical and social implications of this technology and to enable society to produce appropriate long-term policies regarding this technology should the time come when present concerns about safety have been addressed.

V. Finally, because scientific knowledge is essential for all citizens to participate in a full and informed fashion in the governance of our complex society, the Commission recommends that:

· Federal departments and agencies concerned with science should cooperate in seeking out and supporting opportunities to provide information and education to the public in the area of genetics, and on other developments in the biomedical sciences, especially where these affect important cultural practices, values, and beliefs.

This is the unabridged text of chapter 6 of the NBAC report on human cloning (June 1997).

PART V
Fiction and Fantasy

World of Strangers

Lisa Tuttle

My mother and I were not related.

For a long time I never noticed, although our differences were many and obvious. Throughout infancy, she was simply my world, and I had no reason to question it. I didn't know what was normal, and I was not unhappy.

It was only after I went to school and began to look around me and question what I saw that the oddity of my family situation became apparent. I could not help noticing the difference between me and my mother, and the other parents when they came to collect their children from school. Like me, most of the children at my exclusive, very expensive school had single parents. But unlike me, they were all precisely matched. Day after day as I watched the fathers walking away with their miniature selves, the mothers so happy with their identical little daughters, I felt more and more powerfully the strangeness of my situation. Why didn't my mother have a little girl? I was a boy—so where was my father?

I worried that it meant there was something wrong with me. My mother never commented on it; could it be she hadn't noticed? That she didn't know? For a long time I hugged my fears to myself and said nothing, but one day, inevitably, it came bursting out and I demanded, "Am I adopted?"

"Darling Nicky! Of course not! Why should you think such a thing?"

I began cautiously. "Well, I don't look like you."

"Lots of children don't look like their parents."

"Not at *my* school."

My mother might have saved herself some grief—or at least put off the day of reckoning—if she'd simply sent me off to the local school. Among the poor and ordinary the two of us wouldn't have stood out like the freaks we were, and it might have been years before I thought to question my origins.

She shook her head. "Nicky, you're not adopted. You know I'm your mother. You know I gave birth to you—you've seen the video!"

I remembered; I'd watched it often with a sickened fascination. The groaning, crying woman in labor was undoubtedly my mother, so who else could that slimy, bloody baby be but me? I nodded. "Then who's my Dad?"

A little line appeared between her eyes, and her mouth got tight. "Who *have* you been talking to, Nicky? You don't have a father!"

I did know some basic biology, garnered from school, CD-ROMs, and television. I knew about cloning: the nice, modern way of continuing a species. I also knew how animals reproduced, when left to themselves, and that most of the world's human population, through poverty, ignorance, superstition, or sheer perversity, continued to breed in that messy, animal way—but not people like us! No wonder she was offended. But I was offended, too, that she should think I was stupid enough to believe that I'd been cloned like all my friends, when it was so obviously impossible.

"What's the matter?" I taunted, "Did you forget his name? You're not religious and you're not poor, so I guess that leaves stupid. Was I an accident?"

I had never seen such a murderous expression on the face of my usually adoring mother. For a split second I thought she would hit me. And I deserved to be hit. What I suggested was outrageous. I could hardly believe my mother had ever had real-life sex with another person. She certainly hadn't since I was born. Cuddles with me, and a weekly full-body massage from a professional masseur were her only physical intimacies with others. Probably, like most adults, she indulged in virtual sex with the equipment in her bedroom—but at that time of my life I knew nothing of such things.

Then her anger passed, and she stared at me sadly. "You don't know what you're saying. Poor Nicky, has this been worrying you? I'm sorry. . . . I promise you, you were no accident. You were very much planned. And you were cloned."

"Do you think I'm stupid? Look at me! I'm nothing like you! I'm a boy—I have dark, curly hair—there's no way I'm your clone!"

"No—listen to me—you're not *my* clone, but you are. . . . I cloned you from a man—the man I loved—and I carried you in my womb for nine months—"

I didn't want to hear about her womb. "Is he dead?" I demanded. "My original. My father. Is he dead? Is that why you have me?"

"No. I'm not sure. I don't know. Oh, Nicky, you're not really old enough to understand. . . ."

"I am so! You have to tell me—it's not fair—he's my real parent, not you. I should be living with him, not you. You have to tell me why I'm not."

She tried to be firm, but she could never resist me for long when I was determined to have my way, and also, I think, after so many years of secrecy, she needed to share her secret.

"I loved him," she said. "I really loved him. That's the important thing for you to understand; that's why I did it. I would have done anything for him, would have been whatever he wanted—but he didn't want me, that was the problem. He loved sex, and novelty, and the excitement of new beginnings. After about a month he'd had enough of me, he wanted to move on. I wouldn't have stopped him taking other lovers, if that was what it took to make him happy. I told him so. I wouldn't interfere. I'd learn to live on whatever he had to spare, as long as he would spare me something. But he didn't even want to have me around. I bored him, he said."

And she bored me. I didn't care what grown-ups got up to in the name of sexual love; what did that have to do with *me?*

"I can't expect you to understand," she went on. "I hardly understand myself. It all seems so long ago. . . . I was a different person before you were born, Nicky. Being your mother has changed me."

"But you're not my mother," I objected. "We're not related. There's absolutely nothing of you in me at all."

"Oh, Nicky, there's more to motherhood than a genetic link! Of course I'm your mother, in all the ways that matter. I love you. I've always taken care of you. I carried you for nine months before you were born—"

"Any cow could have done that," I pointed out.

"But a cow would have had to been *made* to do it. A cow wouldn't have done it for love, or even have understood what was happening inside her. Nicky, the only reason you exist at all is because I loved you. Because I wanted you to be. Genetically, we're not related, that's true. But I couldn't love you any more if you were my own flesh and blood."

I pestered her for the name of my real parent, but she wouldn't tell me. The discussion was closed.

But the damage had been done. She had told me enough for me to draw two conclusions: My mother was a criminal, and she loved someone else, not me.

If my father was dead, there would be no reason not to tell me his name. I thought that the same would be true if for some, unimaginable reason he had *allowed* her to keep his cloned child. If she had, as I first imagined, reneged on a mutual agreement—if he had hired her womb and she had run away with me once I was safely implanted—I was sure she would not have gotten away with it. My father, with the full force of the law behind him, would have tracked her down and got me back, if not before my birth, then certainly after. I knew, with every ounce of certainty in my body (which was, after all, exactly the same as *his* body), that my father would never give up until he found me, and I was just as certain that my so-called mother would not have been able to cover her tracks well enough to confuse him for long. She wasn't even bright enough to tell me a convincing lie about my origins. She didn't live like a hunted person; I was sure she had never feared pursuit.

Which left only one remaining possibility: that she had cloned a man without his permission or knowledge.

The "why" was a baffling mystery. She said she loved my father, yet she had done the most hateful thing imaginable. Because I couldn't trust what she told me, I set about trying to find answers on my own.

From her personal computer I dredged up traces of old journals, notes she'd written to herself, and, although I still found it hard to comprehend, a picture began to emerge.

I'd be his friend, his helper, his servant—I'd be whatever he wants. I'd change my sex for him if it would help. Since he says it won't, I'll abjure sex. After all, it's little enough. If I can't have him, I don't want anyone else inside my body. He doesn't understand: He thinks love is sex. But I can live on less and find it more. I can sublimate, quite happily. His voice in my ear, the sight of him filling my vision, will thrill and sustain me.

I can't live without him. Somehow, I must have him.

It'll never be over for me, as long as I live. He needn't touch me, if he finds me so distasteful; he doesn't even have to talk to me, not even look at me, if only he'll let me draw near, warm myself at the fire of his being.

I lay in his bed, inhaling the scent of him. It was mingled with that of his latest lover, as were the secretions dried on the sheets, but that doesn't matter. I know what is his well enough to filter out the other.

In the old days they'd have burned me as a witch, I thought, as I gathered up the hairs from his brush and comb. But in the old days they'd have un-

derstood my passion, my love for him, as nobody does today. It doesn't mat-
ter what he does to me or says—I'll always love him.

He ordered me out of his house. He's had the locks changed. He warned me
he'll get an injunction to stop this "harassment."Afraid that I'd never see him
again, afraid I'd never have another chance, I struck him in the face, my fin-
gers curved into claws, raking his flesh. I left a line of raised welts. I drew
blood. He cursed me, but I'd got what I wanted, beneath my nails.

His name was nowhere to be found, only her obsession with
him, the why and the how of her crime. She'd stolen skin cells
from the man she'd been pursuing, then run away and created her
own personal, infant replica.

At that moment of understanding I left behind my babyish love
and need for her. No child, it might be argued, ever asks to be
born, but in my case it was much worse. My mother had not given
life; she had stolen it. From that day, I despised her. This was the
first, profound turning point of my life.

The second came nearly a decade later.

I was living in San Francisco, where I'd gone when I'd left home
because I knew my mother had lived there as a young woman, and
it seemed likely that was where she'd met my father. He might
have moved on, of course, but I had no other leads, and in any case
I needed a place where I could live without interference, where I
could lose myself and survive.

It was the right choice. I looked older than I was and I was quick
to adapt to whatever was expected. Now set free from the pretence
of being *her* child, I could be anyone, just another ordinary stranger,
struggling to get by.

Then one day I walked into the café where I'd been working as
a waiter for the past two months, and saw *myself* sitting at a table,
drinking an espresso.

It was worse than seeing a ghost. I felt as if a bucket of cold water had just been poured down my back. I gulped and shivered.

He looked straight back at me with narrowed eyes. He was trying to look calm, but when I got closer I could see he was trembling all over like a horse restrained from bolting.

I sat down across from him. That took all my nerve. There are legends, although at the time I'd never heard them, that to meet your own double is a harbinger of death. I don't know if he knew that. For what seemed a long time we simply stared at each other. Then he said, "They told me my identical twin was working here. I didn't believe it, but they were right. Who *are* you?"

"Who are *you?*" I spoke angrily, because I already knew he wasn't going to give me the answer I wanted, and for a second I thought he was going to argue. Then he gave in, with an impatient movement of his mouth. "As if you didn't know. Charles Nicholas Weller, the second. Don't ever call me 'Junior.' I go by Chaz. Yourself?"

"Nick."

"How old are you?"

"Eighteen." The reply came automatically. I had been saying it for so long I hardly knew if it was true.

He looked startled. "But you're older than me. I was just sixteen last month. *You* should be 'the second.' Father never mentioned you."

Father. The word burned like a hot coal. How I had longed to be able to say it to someone. For years I had dreamed of this fateful meeting, of coming face to face with my double on the street, in a park or a café, and being able to address him as my father. And now the meeting had happened, and it was all wrong. Not my father, but my father's son, the one he must love best of all. Like the woman who had pretended to be my mother, this boy was in the way, blocking me off from the one relationship that mattered.

"He doesn't know I exist," I said. "I've been looking for him, but I never even knew his name."

"But how——?" Chaz's fear had gone, and he gazed at me in fasci-. nation, mouth hanging open slightly, looking like a boy who's just been told the facts of life.

"It's a long story," I lied. "I'd rather tell it just once, to both of you. I'm sure he'll want to meet me. . . ."

I wasn't sure at all. In all my dreams my father had been a man on his own, aware of something missing from his life and ready for me to step in and fill the gap. But, as he already had a son, what would he want with me?

"Sure," said Chaz, so eagerly that it was obvious he felt too secure in his father's love to fear an interloper. "Sure he will. Wow, isn't this something? Just wait till Dad sees you! Come on, brother!"

We left the café together. I didn't even bother telling anyone there that I'd quit.

Ever since I'd become aware of what was wrong in my life, I'd been fascinated, maybe even obsessed, by the subject of fathers and sons. I'd done a research project on it at school, going beyond the simple biological facts into the deeper cultural and psychological importance of the relationship.

Motherhood, by comparison, is unproblematic. Women have it easy. All throughout history even the poorest, ugliest woman could get a child if she really wanted one—if she was barren, there were usually plenty of unwanted children around to be fostered. These days, of course, she can carry her own clone to term.

Long, long ago, perhaps, women were worshipped for their powers of fertility. The children who issued forth from their wombs were gifts from the gods, or ancestors reborn, and had no connection to the men of the tribe at all. However, once the male contribution to reproduction was recognized, it quickly came to assume the dominant role. Throughout most of recorded history it was generally assumed that a child belonged 100 percent to the father, as if it was entirely his seed, or his clone; the woman was no more than the house in which the little sperm-homunculus could live and grow.

It's no wonder, really, that men in the past treated women so badly. Imagine what it would be like if you couldn't build or buy your own house, if you could only have a home by begging or forcing someone else to take you in? Think of women as the ones who had the houses. Deprivation doesn't make for nice people. The only way a man could ensure that his children got born was by totally possessing a woman, and keeping her locked away from other men who might want to use her to house their own offspring.

Not until 1960—coincidentally around the same time as the development of really effective contraception, and the women's liberation movement—was it incontestably, scientifically proved that every child was created half by the mother, half by the father. As women were finally granted equality with men, in all areas of life, the pendulum swung wildly the other way. Once upon a time, children had belonged wholly to their fathers. If a woman left her husband, even if a divorce was granted, she had to leave her children behind. Even in the case of the father's death, *his* family had more rights over the fate of the children than their mother did. By the 1970s, legislation rapidly changed this, to recognize that children belonged equally to both parents. But when push came to shove, or into a court of law, the woman's half-share was generally given precedence. Men's paternity could be proved beyond a doubt by genetic testing, but rather than giving him the right to his own child, paternity usually did nothing more than establish his monetary responsibility. Equal ownership is a con. It doesn't work. As King Solomon proved long ago, there are no shares in a child; it's all or nothing. And as men in the past instinctively recognized, either a woman owns her own body—and the children who come out of it—or a man does. As the old saying has it, "A man cannot serve two masters." Nor a child two parents.

Human cloning set men—and women—free. We don't have to fight each other any more. Now men have the same, inalienable right as women. We can reproduce ourselves. We can have our own children.

That, more or less, is how I ended my class presentation, all those years ago. I got an A-plus. I still remember the applause from my classmates.

Yet the freedom I proclaimed then was not yet as actual as I made it sound: When Chaz was an infant, men were still dependent on women for the gestation period, although at least use of women in this way could be a relatively straightforward, commercial transaction. The development of a completely satisfactory artificial womb lagged, perhaps because there was no perceived demand. There were always plenty of women willing to rent out their wombs; it wasn't as profitable as regular prostitution, perhaps, but some women preferred it. Another possibility was animal wombs: Dairy cows were fine, and farmers were always on the lookout for extra sources of income. But some studies suggested possible developmental delays, or later psychological problems in the cow-born. A few men were dedicated—or mad—enough to actually give birth to their own children, but the failure rate (including parental, as well as infant, mortality) was significant and the major surgery involved for a one-time event remains daunting to this day.

Charles Nicholas Weller devoted himself to his forthcoming fatherhood like an ancient scholar to his holy task—but he wasn't mad. He had the injections necessary to allow him to nurse his child for up to six months, and he took great care in choosing the woman whose womb he would rent. He paid extra to ensure she was surrounded by positive environmental influences throughout her confinement. He visited her apartment every day in order to speak to his unborn child, carefully constructing a prenatal bond. Birth was, of course, by cesarean section, with the carrier under total anesthetic: She never so much as glimpsed the baby who was delivered into his father's waiting arms, laid on his naked, swollen breast to suck.

I winkled all these details out of Chaz as he drove us out to his father's beautiful bayside home, and I burned inside with the single, jealous thought: That baby could have been me. *Should* have been me.

"Dad!" Chaz shouted as we entered through the glass-walled conservatory at the side of the house. "Hey, Dad, I'm home! You'll never guess what!"

I felt a murderous, infantile rage that encompassed both Chaz and my mother. This was my birthright, stolen from me.

And then Charles Nicholas Weller walked in, and my whole world shifted on its axis. Suddenly, everything changed.

I no longer hated my poor, mad, sad mother, for it seemed she was no longer the wicked witch, but a good fairy in disguise: If she had not kidnapped me, I'd never have known this happy ending.

I looked at Charles Nicholas Weller, my original, and my mouth went dry. I could feel (in my imagination) the tiny hairs lifting on the back of his neck. Of course, he hadn't been expecting me. But neither had I expected my own response. His blue eyes looked into their reflection in mine, and my penis swelled and stiffened. I wanted to fall down and worship at the feet of this god; I wanted to fling myself on this man.

All my envy of Chaz evaporated. I wouldn't have stood in his shoes for all the world. I didn't want to be this beautiful man's son—I wanted to be his lover.

His eyes widened as if he'd read my thoughts. I looked at his trousers and saw his desire before, I think, he knew it himself.

"Chaz," he said, in a strained voice, still looking at me. "What is this? *Who* is this?"

"It's obvious, isn't it?" said Chaz. "But he hasn't said *how*."

I told him my mother's name. This drew a blank at first—it must have been years since he'd thought of her—but then he touched his face, fingers tracing where her nails had raked, the very point of my origin.

"Oh," he said softly. "Son, please believe me, I had no idea. If I had, son—"

"No," I said, softly but emphatically. "I'm *not* your son. I'm yours, but not your *son*. I want you, but not to be my father."

He stared at me for a moment, bewildered, grappling with his feelings—I know, I had the same feelings—and then, gradually, rising through the bafflement in his eyes, I saw his desire. He wanted me. Not as a man wants a child, but as he yearns for his other half, the missing part that will make him ecstatically whole.

Chaz broke into our charged silence, whining like a child. "Hey, what is this? I thought this guy was going to tell us his story?"

"Sure," I said, backing off. "Of course."

"What am I thinking of? We're not very hospitable, keeping you standing out here like this! Come into the family room. Chaz, go get us some drinks, please. What would you like, uh . . . ?"

"Nick," I told him. "Call me Nick. A beer would be great, or a soft drink. Whatever."

It was obvious to both of us what we wanted and what would happen later that night, but we were both grown-ups. We knew how to behave. Charles was careful not to lay a finger on me until Chaz had gone off to his bedroom, and it was as well he was so careful, because the moment his flesh brushed mine we couldn't stop ourselves rushing together in the futile yet strangely satisfying struggle to merge completely.

All our care to tread lightly around Chaz's feelings was pointless. We might just as well have made love in front of him. He hates me, and blames me for his lost happiness. All his life he's had his father to himself, and he obviously imagined him as a sexless being whose whole satisfaction was bound up in his son's happiness. Children don't like to think of their parents as sexual beings.

But Chaz is not a child anymore. He's older than I was when I left home. There's no point in telling him now, because he's still too young to appreciate the long view, but, far from taking anything away from him, my arrival has paved the way for his future happiness.

"It's disgusting!" he hissed at me that first day, when he realized his father and I were lovers. "Shouldn't be allowed! It's incest!"

But of course it's not. Sex between Chaz and Charles would be unthinkable, a corruption of the father-son relationship. But Charles is not my father. Not in a biological sense, and not in a so-cial or cultural sense, as I was raised by someone else in complete ignorance of his existence. It's idiotic to talk about crimes against nature in this context. We're not part of nature, Charles and I; we've transcended it, remade it.

We are the forerunners of a new race. It was for this that I was born. Being so happy, we long to share our happiness with every-one. And we will, but it can't happen overnight. We've already planned for Chaz's future, though: Without his knowledge, Charles has had him cloned, and set aside some safe investments to ensure that the boy is raised in good circumstances, completely anony-mously, by foster parents. Once he's reached sexual maturity he'll be told who he really is, and Chaz will be informed of his existence.

When they meet, nature—our new-formed nature—will take its course. By then, any lingering resentment he may feel towards me will be washed away on the tide of happiness. Someday, after we are all dead, no doubt, everyone will make such arrangements for the happiness of their descendants, and the world of strangers into which I was born will have become a world of lovers.

Flourish Your Heart in This World

Felicia Ackerman

aurel knows she is not supposed to play favorites, but she cannot help liking Mrs. Noll better than the other hospice patients. Mrs. Noll is someone Laurel would like even if it weren't a professional duty, someone Laurel would be happy to spend time with at a party. Now Laurel is sitting at Mrs. Noll's bedside, in the intricately carved teak armchair Mrs. Noll has brought from home. The chair is unupholstered and the wood is slippery, but it is surprisingly comfortable. "Did you get this chair in India?" Laurel asks.

Mrs. Noll nods, smiling. Although she is seventy and dying, she is still vaguely beautiful, with her blue eyes and thick white hair—a marvel in this place where most of the patients (Laurel keeps forgetting to call them guests) have had chemotherapy. "My husband and I got the chair and the rug on our twentieth-anniversary trip to Asia, and that"—she gestures at the framed calligraphy-and-goldleaf Malory quotation on the wall—"was his first-anniversary present to me. It is lovely to be able to have so many of my own things with me here. It makes up for all the psychobabble."

"I know." The homelike atmosphere—you can bring your own furniture, redecorate your own private room, and one woman has even brought her parakeet—is of course a major selling point for this residential hospice. Sometimes this reminds Laurel of the Richard Armour book that says college dormitories have a homelike atmosphere, especially for people who come from homes with hundreds of bedrooms, each of which is occupied by a stranger. But she has to admit that this is a much pleasanter place to die than the hospital where she used to work. That's why she works here now. And there aren't hundreds of bedrooms here, just thirty-four.

"And the coverlet is from Nepal," Mrs. Noll is saying. "I have no regrets about my travels. They were worth it."

"I know," Laurel says again, although she also knows that without the Asian trip, not only these exotic items but Mrs. Noll herself would not be here. "Hepatitis," Mrs. Noll said at the intake interview six weeks ago. "Hardly the worst thing you could pick up in Asia. It takes twenty-five years to destroy your liver, and for most of that time, there are almost no symptoms. And once people learn that your cirrhosis of the liver comes from unhealthy travel rather than unhealthy drinking, they become a lot nicer. Fascinating, isn't it?" she added, winning Laurel's heart on the spot.

Now Mrs. Noll is inching her way up against her silk-covered pillows, inching her way farther into the sunlight that, streaming through the curtain, is making a lacy pattern on her face. "Have you heard anything more from your cousin in California?" she asks.

"I talked to her last night. She says that when you're actually doing it, you don't go around thinking, Isn't this weird, I've gotten myself cloned. You're just pregnant."

"When did they first clone a person, a couple of years ago?"

"A little more. 2003. It's still pretty experimental."

"How very interesting." Mrs. Noll sounds interested, which is part of her charm. Of course, the social workers are paid to be interested in the patients, not the other way around, but why shouldn't you like someone who takes an interest in what you say? And who often has a box of chocolates invitingly open on her night table—Mrs. Noll no longer eats chocolates, but she likes to watch other people enjoy them. Laurel takes a chocolate-covered cherry, feeling the slithery sweetness spread through her mouth. She is about to say that Juliana, who has always gotten everything she wanted except a baby, is now getting to have the baby too. But Mrs. Noll is saying, "I suppose I'm a hopeless reactionary. But I still think it's unnatural. Like going to a Mercicenter instead of a hospice. Cloning, suicide—people have to control everything nowadays. At my age I figure I have the right to be a reactionary if I want to be."

"You certainly do," Laurel says.

"I wanted children at first, but my husband and I grew to be very glad it was just the two of us."

"Juliana spent eight years trying to get pregnant," says Laurel. "I don't think she would ever have gotten to be glad. She's not much for adjustment. She's used to getting what she wants."

"I believe in taking life as it comes," Mrs. Noll says placidly. "Hardly a fashionable attitude for a woman these days. Unless she's seventy and dying in a hospice," she adds, her tone without rancor.

Laurel feels ridiculous saying "I know" for the third time, but in fact she does know. Seeing Mrs. Noll here, dying serenely but with flair, makes Laurel recognize the insidious contagiousness of the hospice mentality. How lovely it would be, she often thinks, how lovely to stop struggling to find a man who could love a woman who never had much verve and is now thirty-six and fading, to stop hoping for a stroke of good fortune to change her life, just to accept, accept, lie back in a perpetual warm bubble bath of acceptance. But only part of her feels that way. The other part can't bear to abandon her dreams.

"You don't like your cousin, do you?" Mrs. Noll's voice is weakening; her eyes are closing.

"Well," says Laurel, "I got off to a bad start. Juliana lived across the street from me until I was eighteen, and she had a bigger house, a bigger wardrobe, a swimming pool in her backyard, and a red BMW. Would you like someone like that?"

"As a teenager?" Mrs. Noll murmurs. "I doubt it."

"And all those fancy trappings couldn't hide the fact that underneath it all, she was prettier and smarter than I was."

Mrs. Noll's eyes snap open, as if she has received a sudden infusion of energy. "I suppose nowadays they would say you should sign up for one of those self-image therapy weekends."

"So I can think I'm as pretty and smart as my Stanford-physics-professor cousin who has blond hair she can sit on?"

Mrs. Noll laughs, thrilling Laurel with how momentarily vigorous she sounds. "Oh, you are delightful," Mrs. Noll says, "especially as compared with the general run of people around here."

Laurel feels as if she has stepped into a pool of sunlight.

"And you can be attractive." Mrs. Noll's voice, kind as ever, has taken on an appraising tone. "All you need is lipstick and just a bit of moss-green eyeshadow and maybe blusher. You should also start wearing richer colors, to bring out the rich auburn highlights in your hair."

"What rich auburn highlights in my hair?"

"The highlights you will have when you start using an auburn rinse, which is something else you should be doing."

Laurel is actually considering following this advice. She puts her hand on Mrs. Noll's shoulder, recalling how last week Mrs. Noll said, "When you had thirty-eight years of marriage to a wonderful husband, the last thing you want is to have your hand held by a hired professional, but of course, it's different with you, dear."

A moment later there is a knock on the door, and then Ellen Lefferts, the hospice director, is walking into the room, pulling up a chair, and sitting down by the bed. "I hear you had a bad night," Ellen says.

"Only for a bit," says Mrs. Noll. "Nicole came and gave me a shot right away. The symptom relief here is very effective."

"That's what we're here for," Ellen says.

"Thank you."

"I also stopped by to invite you to our support group that will be meeting at three in the solarium." Ellen leans forward; she is wearing a blue cotton dress with sprigs of coral flowers.

"No, thank you."

·

"Are you sure?"

Mrs. Noll nods.

"I hope you will let us know as soon as you are ready to talk about your feelings. We're caring for the complete you."

"Thank you."

As she is leaving, Ellen asks Laurel to stop by the office before going home for the day.

"How I wish there were some tactful way to get it across to her that while I'm quite happy to die serenely, like a good hospice poster child, I don't want to talk about it," Mrs. Noll says, almost as soon as Ellen has left and closed the door behind her. "It's not an interesting subject. Besides, some things are private."

"We're caring for the complete you," Laurel says with a grin.

"Oh, well," says Mrs. Noll. "I suppose we shouldn't be too hard on her. When I think of how terrible I felt when I came here and how comfortable this place has managed to make me, I'm ready to forgive her anything. Well, almost anything. I hope she isn't giving you trouble for spending too much time with me."

"Oh, no," Laurel lies.

"Good. Let me see your hands."

"My hands?"

"I had thought perhaps some coral nail polish. The flowers on her dress gave me the idea." All at once, Mrs. Noll looks exhausted. She is sinking back against the pillows, but she manages to say, "We're caring for the complete you," before drifting off to sleep.

*   *   *

Ellen's office is full of plants. There is a thicket of pink begonias on the mantelpiece and an herb garden on the windowsill. There are two African violets in green flowerpots on the desk. A potted avocado tree is threatening to burst through the ceiling. Plants are the main decoration throughout the hospice building. Maybe Ellen has read that plants symbolize the cycle of nature, the passing of the old and the coming of the new. Or maybe she just likes plants. She is certainly good at growing them; the thicket of begonias is the most luxuriant Laurel has ever seen. Above it, a placard proclaims, "Medicine should be high-touch, not high-tech." Laurel figures this is not the place to mention Juliana.

"Did Minnie talk to you yesterday about the party?" Ellen is asking.

Minnie is Mrs. Noll's cousin and her only visiting relative, a retired nurse who began visiting recently and comes on Thursdays, dressed in tennis whites on her way to the courts. This strikes Laurel as tactless, but Mrs. Noll doesn't seem to mind, although she privately calls Minnie "Minnie Mouse." Minnie even looks a trifle mouselike, with her bright eyes, sleek gray hair, and perky features. "What party?" Laurel says.

Ellen rests her elbows on her oak desk, steeples her hands, and explains that Minnie will be having a sixtieth birthday party a week from Sunday. Most of the extended family will be there. It will be an opportunity for Mrs. Noll to reconcile with relatives she's been estranged from for years. But she is resisting the idea of going. "If you would talk with her and try to get at the root of the problem."

"Well," says Laurel, "if she doesn't want to go, I don't see what there is to talk about."

Ellen sharpens the steeple and reminds Laurel that patient plus family is the unit of hospice care.

"Mrs. Noll is dying." Laurel turns her head, with the result that she is staring into an African violet. "I'm not going to push her to do something she doesn't want to do just to make her relatives happy."

"It's not just to make her relatives happy." Ellen has unsteepled her hands, which now lie flat on her desk. "People die more peacefully when they are at peace with their families."

If Mrs. Noll were any more peaceful, you could package her and sell her as a tranquilizer, Laurel is tempted to say. Then it occurs to her that Mrs. Noll might enjoy a new opportunity to poke mild fun at Ellen's ideas about good adjustment. "Okay," she says. "I'll talk to her. I'll do it."

*   *   *

"You did it!" says Mrs. Noll when Laurel comes in after the weekend.

"Do you like it?"

"The hair is perfect and the lipstick is fine too. But the eyeshadow is all wrong. Oh, I don't blame you, dear. I blame myself. I should have seen that you needed golden brown, to complement your new auburn hair."

"How about this?" Laurel opens her purse and takes out her little palette of eyeshadows—mauve, two greens, and mahogany.

A few minutes later, she is looking into Mrs. Noll's silver-backed mirror at her new mahogany eyelids and, since Mrs. Noll has gotten her to intensify the lipstick, brighter lips. Still no competition for Juliana, and maybe the effect is rather conspicuous, but why not?

"Tell me what you've been up to, aside from a makeover for the complete you," says Mrs. Noll.

"Well, I promised Ellen I would talk to you about Minnie's birthday party."

"What?"

"She thinks you will be more at peace if you are at peace with your family."

"Tell her if I were any more at peace, I'd be dead already."

Laurel looks at her hands. The coral nail polish has already begun to chip. "Juliana's invited me to visit in about seven weeks, after the baby is born."

"I thought you didn't like her."

"I can't really dislike her. She's too nice. Anyway, a friend of hers will be at Stanford for a mini-course. She thinks he and I might like each other. She wants to—"

"That's marvelous."

"I'm not sure I want to go." But then Laurel realizes Mrs. Noll might suspect the reason—Laurel doesn't want to risk being away when Mrs. Noll dies—so, keeping her eyes focused on the embroidered Nepalese coverlet, she launches into an account of the contagiousness of the hospice outlook and how she seems to be losing what little get-up-and-go spirit she ever had. "Not that I'd tell Ellen that," she adds. "She's always talking about how the terminally ill can be an inspiration to us all. But I doubt this is the kind of inspiration she has in mind."

Laurel looks up, expecting Mrs. Noll to be amused. But Mrs. Noll looks horrified. "You mustn't—all this acceptance is fine for a dying old lady like me whose husband is dead and whose life is over. But a young person like you who has never had an abiding love . . . You must take this opportunity, Laurel, flourish your heart in this world."

"What?"

"Flourish your heart in this world. It's from the Malory passage on the wall. Get up and read it. You might as well become familiar with it. I have left it to you in my will."

Laurel's eyes fill with tears.

"Read it, dear," Mrs. Noll says gently. "And you must try not to cry when you are wearing eye makeup."

Laurel walks over to the wall. "Therefore, like as May month flowereth and flourisheth in many gardens, so in likewise let every man of worship flourish his heart in this world," she reads aloud after a moment.

" 'Worship' meant honor in the fifteenth century. A man of worship was an honorable man. Won't that be a nice thing to hang in your living room?"

Laurel, working her way through the difficult calligraphy and unfamiliar phrasing, starts to giggle shakily.

"What's so funny?"

"But the old love was not so; men and women could love together seven years, and no licours lusts were between them, and then was love, truth, and faithfulness," Laurel reads aloud.

" 'Licours lusts' meant sexual pleasures," says Mrs. Noll. "What is funny about that?"

Laurel stifles a mental image of a man walking into her living room and seeing a wall-hanging that says he should go seven years without lusts. She turns around. Mrs. Noll is smiling, her eyes half-closed. Probably the passage reminds her of her husband, although Laurel doubts they waited seven years to have sex.

"I guess you don't read fashion magazines, with their articles about how long to wait before sleeping with a new man," she says, sitting back down at Mrs. Noll's bedside. "On the third date, or the fourth, or if you're really conservative, maybe wait a few months. I don't recall any of them saying seven years. But of course I'll put it up in my living room. I want to. I can't tell you how much I . . ." She swallows hard.

"Perhaps this will help you attract a better sort of man. At any rate," Mrs. Noll says sleepily, "the main thing for now is to get you to California. I'll go to Minnie's birthday party if you'll go to Juliana. I'll visit my cousin if you'll visit yours. Is it a deal?"

"Yes," says Laurel. "It's a deal."

*   *   *

Her side of the deal didn't turn out so badly, Mrs. Noll tells Laurel on Tuesday of the following week. "They were all terribly sweet to their dying old relative, partly out of pity and no doubt partly because they're all hoping to be remembered in my will."

Laurel glances out the window. The day is bright and blue, with pink and yellow flowers swaying in the hospice garden, the kind of day that would bring hope to anyone, even a dying old lady and an aging unloved social worker.

"I'm going to have some fun," Mrs. Noll continues. "People who are after a dying old woman's money deserve to be toyed with. That's why I agreed to stay through yesterday so I could spend more time with some relatives I hadn't seen in decades. Minnie's also invited me to a luncheon this Thursday for her granddaughter, and I agreed to go if I feel up to it. And if I'm being unfair and all she wants is the pleasure of my company," Mrs. Noll's voice is fading, but her expression is still alert, "well, that's all she's going to get. Have you made your plane reservations for California? Have you practiced saying, 'Oh, she looks just like you'?"

*   *   *

"Oh, she looks just like you," Laurel is saying six weeks later. "You'd better not commit a crime tomorrow, or she might get arrested."

Juliana beams, leaning back in her lawn chair, and runs a finger down the baby's round pink cheek. Juliana's cheekbones are high. Her hair is golden, and the baby is almost bald. Ten-day-old Beatrice Parker-Denison looks about as much like her mother as like Laurel. But if genes settle the matter, in thirty-six years Beatrice will have her mother's shining hair and perfect cheekbones. And Juliana will be seventy-two. Maybe still vaguely beautiful, like Mrs. Noll, but nothing to compare with the youthful version. Laurel has heard people say you have to be very egotistical to get yourself cloned. Now it strikes her the opposite is true. Generous and noncompetitive, that's what you have to be—imagine bringing someone into the world who will be genetically just like you, but as you age, decades younger.

Paul, Juliana's husband, is standing behind her lawn chair and dangling a red ball on a string above Beatrice, the way Laurel earlier saw him dangle a toy in front of the family cat. "Look at this," he says. "A family in their backyard, what could be more natural? I wish that patient of yours who says cloning is unnatural could see this, Laurel."

"Everything that is usual appears natural. That's what Vicky always said," says Dan, opening a can of beer. "Of course, she got it from John Stuart Mill."

Laurel tries to catch Juliana's eye, but Juliana is blissfully absorbed in the baby. As absorbed as Dan apparently is in Vicky. But at least Juliana's daughter is alive. Dan's wife Vicky has been dead for seven years, but except for the past tense, he talks about her as if she has just gone out to the supermarket. Vicky liked teaching middle school because no matter how ruthless the pecking order was, the girls were so young you could always tell yourself that someday the last would be first. Vicky didn't like wine and wasn't interested in learning to like it; Vicky said she worked at her work, she wasn't going to work at her fun. Vicky once had a pet swan. Laurel thinks she would have liked Vicky. But she can't imagine how she is supposed to attract a man who still sees himself as Vicky's husband.

"Vicky—" Dan begins again.

Beatrice shrieks. Good for you, Laurel says silently. "Excuse me," says Juliana and gets up, carrying Beatrice into the house.

Laurel hesitates, then follows. Juliana is sitting on a loveseat in the living room, a towel draped over her chest, breast-feeding Beatrice. Laurel is surprised. She had a vague idea that in California breast-feeding would be as public as politics.

"So what do you think?" Juliana asks.

"She's beautiful."

"I mean about Dan."

"Dan?" Laurel sits down in a green velvet armchair that turns out to be even softer than it looks. "I think if he gets involved with me, he'll be committing adultery. Why are you trying to fix me up with someone who's still in love with his wife?"

"Widowers who were happily married make ideal husbands."

"No, they *made* ideal husbands. If widowers want to get married again, it means they love marriage, not their wives. True love," Laurel picks up a pretzel stick from a bowl on the coffee table and holds it aloft like a scepter, "does not look for replacements."

Juliana giggles. "What've you been reading, *Riveting Romances?*"

"Malory. Mrs. Noll has got me reading him. People in Malory's world don't look for replacements."

"You can't really believe—"

"I really believe Dan isn't looking for a replacement. Neither did Mrs. Noll. She wouldn't be so serene about dying if her husband were still alive." Laurel gazes straight ahead. The opposite wall has a mural of wildflowers. "Too bad Vicky died too long ago for Dan to clone her."

"So he could marry an infant? Besides, he wants Vicky, not a clone."

"Then why do you think he'd want me?"

"You have a lot in common. You're both so romantic. No one can mourn forever."

Oh yes, they can, Laurel wants to say. But Beatrice is falling asleep, and Juliana also seems to be shutting down, her face at once peaceful and exhilarated as she looks at her daughter, as though falling in love and assured of reciprocation. Mrs. Noll grew to be happy to be childless, happy to flourish her heart entirely unto her husband. Probably Dan, who is also childless, was the same way. Probably he was as wonderful a husband as Mrs. Noll's was. If only . . . Laurel is tired too, jet-lagged. She feels as if she is sinking into the field of flowers on the wall. She can almost smell the flowers; then she realizes there is a vase of roses on the end table. Juliana's plushy cat, blue-gray like Dan's eyes, is rubbing against Laurel's ankle. How soft. Everything seems to be conspiring to make her fall in love. But she doesn't want to be like the women in Malory's world who fall in love with Lancelot because he is so devoted to Guinevere and they want all that devotion for themselves. They can't see that going after someone because of the stability he shows in his devotion to someone else is a losing proposition. If he shifts his devotion to you, he no longer has the stability that attracted you in the first place. Anyway, she barely knows Dan.

Laurel has an impulse to telephone Mrs. Noll and find out how she is, but it is too late and she is too sleepy.

*   *   *

The next morning Laurel awakens with an urge to giggle. Imagine, she was practically ready to fall in love with a total stranger just because he still loved his dead wife and Juliana's living room smelled of roses. That's what reading Malory will do to you, she envisions Mrs. Noll saying. Why not give her a chance to say it right now? And why not give her a chance to hear that Juliana's daughter doesn't look like an unnatural clone, she looks like a baby? Laurel picks up the receiver to call Mrs. Noll, only to be interrupted by the cat, who has come into the guest room, bounded onto the bed, and is now inserting her velvety blue-gray body between Laurel's mouth and the receiver, purring loudly. Three times Laurel pushes her away, but the cat keeps returning like a velvet boomerang. Laurel is so absorbed in trying to keep her mouth at the receiver that it is a while before it strikes her that the rings are not being answered. She hangs up, tries again, waits ten minutes, and tries a third time.

Maybe she's just out of her room, Laurel tells herself, don't get upset. But on the other hand, why not? People in Malory's world get upset all the time. They don't worry about being well adjusted. And Mrs. Noll is rarely out of her room anymore. In the past month, she has gotten weaker, although the hospice doctor predicts a couple of months more for her. He has admitted he can't be sure. Laurel will have to sound calm if she is going to call Ellen. Ellen thinks Laurel is too involved with Mrs. Noll, just as most people would think Dan is too involved with the memory of his dead wife. Where do they keep the rule book? Laurel picks up the receiver again, concentrating so hard on how calm she is going to sound that it is not until the telephone starts to ring that she realizes the cat is still on the bed.

"Ellen Lefferts."

"Hello. This is Laurel." She tries to push the cat away.

"This is a terrible connection. Do you hear buzzing on the line?"

"It's just a ca—Yes, it's a terrible connection." The purrs are rising like tidal waves. Laurel scoops up the cat and plops her on the floor. "Is Mrs. Noll all right?"

"I have such good news." The good news is that Mrs. Noll has gone back to Minnie's for a few days. Minnie is a retired nurse, re-member? Isn't Laurel glad she persuaded Mrs. Noll to go to the birthday party? Isn't it wonderful when terminal illness leads to family reconciliation?

Laurel gazes out the guest room window, which overlooks the backyard. Juliana is lying in a hammock, with Beatrice on her stomach. Juliana's husband is pushing the hammock back and forth. And in the room, the cat is purring the loudest purrs Laurel has ever heard.

"Yes," she says, "wonderful."

*   *   *

And if Merlin could have enchanted him into believing I was Vicky, we would have lived happily ever after, Laurel is rehearsing in her mind eight days later as she knocks on the door of Mrs. Noll's room, her first stop in the hospice on the day after returning from California. Knock before entering; that's a rule of what Ellen calls hospice philosophy. Give our guests the courtesy you would give a guest in your own home.

"Come in," says a man's voice.

Laurel's first thought is that it must be the hospice doctor. But the doctor would not be lying in the bed instead of Mrs. Noll. He would not be the reason the room now has none of Mrs. Noll's pos-sessions, no teak armchair, no embroidered Nepalese coverlet, no Asian rug. No framed Malory passage. And the doctor is not an el-derly black man with a fringe of white hair and a benign pedagog-ical expression, as if he has been reading fables to children.

"Good morning," says the man. "You must be Laurel."

*   *   *

"She's as well as can be expected. She has gone to live with Minnie. She'll be there until the end, so she can die surrounded by her loved ones." Sitting under a hanging basket of ferns that is a new addition to her office, Ellen sounds as pleased as if Mrs. Noll has been cured and is off for a trip around the world.

Laurel presses her palms together. First the jet lag, then the fear, and now this. "I can keep on as her social worker. I want to. I don't mind the extra work."

"She won't be using home hospice care. The family wants to take care of her themselves." Ellen steeples her hands. "Minnie used to be a nurse, you know."

*   *   *

Of course, it is only because she felt so close to Mrs. Noll that Laurel feels so bereft now. Patients have chosen to leave this residential hospice before. Some go home to die. Occasionally people even leave because they have a remission, although Laurel can count on her eyeballs the number of times this has happened. Then there are the ones they're not supposed to think about, the ones who leave the hospice to go to Mercicenters, pleasant facilities like hospices, except that at the end of a lovely day or two, you get free poison with your tea. Or they'll bring it to your home. Like hospices, Mercicenters have home care for people who prefer to die at home. Our role is to make our guests comfortable enough with *us* that they don't feel the need to turn to *them,* Ellen likes to say.

Mrs. Noll would never turn to *them,* would she? She scorns the idea. She thinks it's unnatural. But if the nausea got out of control . . . Laurel forces the thought away, but it keeps moving in and out of her consciousness, like a floater drifting across her visual field, until finally, as soon as she gets home, she telephones Minnie.

Ten rings. Twelve rings. No answer. Three occurences of this pattern in half an hour, and it's like being back in Juliana's guest room except for the absence of purrs. Don't get upset. Well, why not? People in Malory's world . . . Eventually, at a quarter after nine, Minnie answers the telephone, and Laurel launches into her prepared speech. "Is this Minnie Larson? This is Laurel from the hospice. I was wondering how Mrs. Noll is doing."

"She's doing as well as can be expected," Minnie says.

"I wonder if I could say hello to her?" Laurel draws a pair of concentric circles on her notepad.

"I'm afraid she's past that. She's quite disoriented. But at least she isn't suffering. We're keeping her comfortable."

Laurel's throat feels scraped. "Maybe I could come and visit?"

"I'm afraid she wouldn't recognize you, dear. But thank you so much for calling. She liked you very much, you know."

Not until several minutes after hanging up does it occur to Laurel to wonder why, if Mrs. Noll is so disoriented, she was left alone in the house for over three hours.

*   *   *

To find out if a particular person has died recently in your city, you can log on to the department of vital statistics at City Hall. Laurel remembers this from a murder mystery she read last month. But Mrs. Noll is not listed. So she hasn't died at a Mercicenter, with Minnie tactfully concealing it from the hospice. What else might Minnie be concealing? You could hardly expect to find Mrs. Noll's death listed if Minnie is concealing it in order to get her Social Security checks. But it wasn't her Social Security checks that Mrs. Noll said Minnie was interested in. It was her will. What if Minnie has gotten Mrs. Noll to make a will in her favor, and now Mrs. Noll is bound and drugged? Or maybe Mrs. Noll has become so weak and disoriented that Minnie has no need of rope or drugs to feel safe leaving her home alone. Or maybe Laurel reads too many murder mysteries. Maybe Minnie is taking perfectly good care of Mrs. Noll and just had her telephone unplugged for a few hours so as not to be disturbed.

All through the following morning, through the staff meeting and the session with the gentlemanly new occupant of Mrs. Noll's room, Laurel is making plans. At lunchtime she telephones Minnie. No answer. Then she drives to a jewelry store, where she buys a pair of garnet earrings she can always keep for herself if no one answers Minnie's door.

"I found an earring wedged behind the night-table drawer. I figured it must be Mrs. Noll's," Laurel rehearses in her mind fifteen minutes later as she pulls up in front of Minnie's house. The curtains are closed, a magazine sticks out of the mailbox, and no car is in the driveway. She walks down the marigold-bordered path and rings the doorbell. She waits five minutes, rings again, waits another five minutes, then bangs the knocker as loudly as she can. She puts her ear to the door. Still no sound from within. Of course, Mrs. Noll may be unable to scream. But Laurel is beginning to suspect there is no one inside the house.

*   *   *

"Maybe Minnie was taking a bath," Juliana is saying late that evening.

"And yesterday she just happened to have her phone unplugged until nine-fifteen?" Laurel shifts the receiver to her other ear. "Look, I went back today after dark, and still no one answered the door, and the porch light was on but the house was all dark inside. Even in back—I drove around the block. I think Mrs. Noll isn't there anymore."

"But—"

"Juliana, could you do something about the cat? I can hardly hear you." Laurel takes a sip of water. "I can't talk to Ellen," she continues after Juliana has removed the cat. "She'd probably say I need to go to a support group for stressed hospice professionals."

"Maybe you do. But that won't help you find out whether your suspicions are right. What you need," says Juliana, "is to go to the police."

Laurel stares at her hands. In memory of Mrs. Noll, she is still wearing coral nail polish. In memory of? Where did that come from? "I don't know if they'd get involved. Anyway, I'd prefer something less official and more discreet. So if it turns out to be nothing, Minnie won't have to know. Maybe I'll investigate a little more on my own."

"Be careful," Juliana says.

*   *   *

Three times during the next week, Laurel takes one of the earrings to Minnie's on her lunch hour. She drives past the house three evenings after dark. No one answers the doorbell at lunchtime, and in the evening there is never light in more than one place. Surely Minnie is out during the day and alone here at night? The vital statistics department still has no listing of Mrs. Noll's death. Laurel is now desperate enough to call the police department of missing persons. Mrs. Noll, however, does not qualify as a missing person just because she is missing to Laurel.

"She's seventy and dying," Laurel protests.

"There's no evidence of foul play. Nothing for us to investigate. Even dying seventy-year-olds are entitled to their privacy."

Laurel has to admit that the last part sounds like something she might say to Ellen. But that Saturday, after calling Juliana, describing what she is about to do, and ending, "If I don't call you back today, call the police," she drives over to Minnie's and rings the doorbell.

Minnie looks so different in her old slacks, T-shirt, and no makeup that Laurel probably would not have recognized her on the street. Apparently, it's mutual. Minnie is blinking and saying, "Yes?" in a pleasant but puzzled way.

"I'm Laurel. The hospice social worker. I think Mrs. Noll left an earring. . . ."

How odd Minnie's expression looks, strained and almost pitying. Pinpoints of fear rise within Laurel like a fireworks display. "Is she alive?" she asks abruptly.

"Come in." Minnie steps back from the door. "I suppose I might as well tell you now."

"When did she die?" Laurel whispers.

"She is not dead."

Laurel does not remember walking through the hall and into the living room, but now she is seated on a sofa opposite a tapestry wall-hanging. "Is she still here?"

"She never was here," Minnie says, "and she's not my cousin."

*   *   *

Laurel does not try to sort out her thoughts until she is walking back down the marigold-bordered path. First comes relief, relief plus elation, because Mrs. Noll is not dead. Is maybe not even dying anymore. Is flourishing her heart, not to mention her liver, in this world. Laurel has a wild urge to laugh; then the second reaction sets in. Betrayal. The sweet old lady with her old-fashioned ideas about natural and unnatural, the sweet old lady serenely dying in a hospice, was a fake. What else was a fake? All that warmth and interest—did Mrs. Noll come to see Laurel just as a dupe? Laurel could hardly ask Minnie that, but she did inquire about Mrs. Noll's husband.

"Yes, she loved him very much," Minnie said, as she poured iced tea into Laurel's glass, "but that doesn't mean she was ready to die. You hospice people are all so apt to believe terminally ill people are ready to die."

Only the ones who come to us, Laurel answers silently now as she gets into her car. They don't have to come to us if they don't want to. But Mrs. Noll did want to, at first. Laurel turns on the radio to an oldies station. "Stop! In the name of love," floats into the car. Laurel cannot stop. She cannot stop thinking about how Mrs. Noll came to the hospice so weak and nauseated that she really was ready to die. But after a few weeks of the comfort care the hospice is so proud of, she felt better, so much better that she began having second thoughts. And when Laurel started to talk about Juliana . . .

Laurel turns the corner, drives several blocks, then turns another corner. She is driving aimlessly. Take the adventure, Malory says. Mrs. Noll took the adventure. Like most of the patients, she had her own telephone. No one in the hospice ever knew what calls she made. So she called the university medical center, a hundred and thirty miles away, just to see. Then she called her old friend Minnie Larson, not a retired nurse at all, but a part-time accountant.

How Laurel fell for Mrs. Noll's sympathetic interest in her and in Juliana! But Mrs. Noll was making plans. For a long time she had been ready to die; she hadn't been looking into last-ditch experimental treatments. But the university medical center staff told Mrs. Noll there was a brand-new possibility, a new kind of transplant, still experimental, but with no worries about organ rejection or waiting lists for scarce livers where a seventy-year-old would rank near the bottom. The procedure involved cloning her and making the cloned embryo cells turn into liver cells instead of developing into a fetus. So in a way the cloned embryo was sacrificed. So was the truth. You can't stay in a hospice if you're awaiting a transplant. Laurel knows this. It's right in the rules. Hospice philosophy means palliative care. If you want to take a chance on a life-extending experimental treatment, go somewhere else. Ellen would not have dumped Mrs. Noll on the street, of course. But there would have been a transfer to a nursing home for the six weeks between the cloning and the transplant, and who could be confident of finding a space in such a pleasant facility on such short notice? The hospice is so comfortable, and the symptom relief is so good.

"You have to understand, her life and her comfort were at stake," Minnie said, taking a gingersnap. "You know how important comfort is when you're so ill. She had no other choice."

No other choice but this elaborate deception? Well, Minnie admitted, looking embarrassed for the first time, perhaps it didn't have to be quite so elaborate. But deception is hard to keep in bounds. Hasn't Laurel ever found that? Minnie conceded that perhaps Mrs. Noll went too far, perhaps she even started to enjoy creating her own little world. But she needed an excuse for the preliminary sessions at the university hospital while they ran tests, evaluated her case, and finally harvested the cells from inside her cheek. That was what Mrs. Noll was up to when the hospice staff, so eager to facilitate a family reconciliation, thought she had gone to Minnie's birthday party and then to the luncheon for Minnie's granddaughter. Ironic, Minnie added, because although new life was being created, there would be no birth. Of course, a lot of people would think that is terrible, but Minnie and Mrs. Noll both think *they're* terrible, especially the ones who think it is all right to have an abortion because you don't want to have a child, but not to produce an embryo on purpose because you want to try to save your own life. And what did Laurel think? Minnie asked.

But Laurel, sitting stiffly on Minnie's sofa, was not thinking about embryos at all. She was thinking about deception and betrayal. She is thinking about that now as she finds herself driving alongside a golf course she has never seen before; she is out of the city by now. First relief, then knowledge of betrayal—where has she previously encountered this pattern? In stories about wives, of course, wives who are afraid something dreadful has happened when their husbands don't come home one night, and it turns out something dreadful has happened, but not to the husbands. To the wives. The husbands are happily with other women. Laurel's face feels singed. How unwholesome Ellen would find all this, perfect proof that Laurel is too involved and needs help whether she wants it or not. But this thought raises Laurel's spirits. It makes her feel unconventional and daring, instead of pathetic. Who is Ellen to say whom Laurel should flourish her heart unto? Laurel turns the car again, and soon she is driving along a country road beside a field of wildflowers like the mural in Juliana's living room, like the May in

Malory's world, when hearts begin to blossom and to bring forth fruit. Laurel's heart is beginning to bring forth fruit right now. Why shouldn't Mrs. Noll grasp at a second chance at life? And why suppose she was using Laurel? Maybe she was trying to protect Laurel by not letting her in on the deception. Maybe she was afraid Laurel would lose her job if she went along with the deception and Ellen found out. Maybe she was right. Laurel's hands are gripping the steering wheel; she sees her coral fingernails—a memorial, no, a tribute, to Mrs. Noll. Surely all that warmth and interest was no fake. All at once, Laurel has the thrill of relief again, but now it is relief plus triumph, as if she and Mrs. Noll have carried off the deception together. Laurel's talk of Juliana was what got Mrs. Noll thinking about cloning, after all. Surely she would be delighted to see Laurel. Laurel does not have to be back at the hospice until Monday. She can drive to the university tomorrow and appear in Mrs. Noll's hospital room with an armful of flowers. Maybe she should take along her copy of Malory, but on second thought, that won't be necessary. Mrs. Noll has undoubtedly brought her own.

My Clone

C. K. Williams

B row still fused to brow, brain to brain, one of us, I can't tell which, is crying; then an anguishing *release*. This is how my clone comes to me; not elaborated in some mute womb, but wrenched entire from my matter and my mind.

Because what my clone means, at the end, is doubling, multiplying, self-replication, as a person or a species; all its science finally tends towards that, our desire is that, our fear is that.

The first thing I imagine with him is my fear. To see oneself naked before one's naked self. The human dream, the human dread.

The next thing I imagine is his going, his dying, his suicide. He kills himself, does away with what he is, because I so oppress him; he can bear no longer my oppression, bear no more the "life" I've given him.

He kills himself because he knows that what I ask of him is irresolvable. To be as wounded as I am, as scarred: to have suffered all I have, just as I have.

I hear him telling me I didn't really want a thing of flesh and blood. That I'd really dreamed of something of a substance other than myself, generated from the tissues of imagination, from an unsmelted crystal of my unconscious. Something from a myth: centaur, minotaur, half-dream, half-human, which would answer all my ancient quandaries. Oedipus, or angel.

Perhaps he was right, perhaps I did feel something like regret, and fear again, when I reached out to him and my hand passed through him.

But still, to kill himself? Yet I know why. Because I ask of him: to be more serious than I am, more courageous, potent, virtuous. Especially virtuous, because he doesn't have the wounds I have to keep him from being so.

I imagine that he'll say to me: then you worried whether I'd be sensitive enough to even be as virtuous as you. You wondered whether human good might be too related to your pathological accumulation of pain. Your hoarding of your wrongs. He'll say: perhaps you didn't *want* me to be "good"; if I were, I might disprove too much.

The inward scars, the failures, losses, disappointments that slash the lining of the heart. How, without them, would he have been enough like me to *indicate* enough, to *mean* enough?

Your fear, he'd say, that I'd *absorb* you, that I'd steal that tiny quantum of eternity which you conceive is yours. Your fear that I'd *replace* you, with no one ever suspecting. How can I exist when what I represent to you is worse than death?

The wounds of pastness, the wounds of never being ready for the wounding, wound of wondering or of wishing for the end of wounding.

I imagine I become afraid that he'd be *more* than me, just as much as ever me, but more focused, less susceptible to all the flinchings and feintings which are the residue of all I've lived. With greater *ardor,* he'd exist, with greater *force.* He wouldn't be bedeviled by my prudence, wouldn't have to flail through all my apprehensions.

How wearisome, he says, your dialectics are. These lurches out to-wards knowledge, these tremblings back to ignorance and fear.

Perhaps my fear is grounded in the way he fits my most intense de-sires: to have a matrix for the self that absolutely matches self. That everlasting human passion to have family, nature, culture, earth, precisely made to fit me, *bent* to fit me, *crushed* to fit me.

Your rule of being, he'd say: is me and me and me. Like-minded, like-bodied, even like-divinitied.

What he might represent is the possibility of that impossibility: in his likeness to me, he'd embody how malignant our ambitions are.

And the way you change your mind ten times a minute. How was I supposed to learn to split myself in half, or ten, to have the wars against myself you want consciousness to be? The way, he'd say, you face forward every moment with ever gleaming hopes for better outcomes to your striving. Don't you, in your mind, he'd say, re-produce one self after another, each you think better than the last? And this hypothetically improved identity which you cast out be-fore yourself: surely, at least sometimes, don't you become him? Aren't you then, in your mind at least, your own multiplying dou-ble, *your own clone?*

I could never clear my conscience of my feeling he was more *dis-posable* than I am. That finally it was this he'd been created for. Is it this that would torment him?

Your fear of mindless masses of me, he'd tell me, our eyes fixed, shoulders rigid. Our duplicability, replaceability: automatons, all primed to cast ourselves into oblivion for masters who have some-how gained control of us, who would steal our minds, our labor, bodies, lives. How dangerous those whose lives are stolen. Won't they do any evil? Toil and war, deception and control. Another of your agonizing fears, he'd say, but what else is human history? Mas-ter, slave. War and toil.

But I *need* you, I might tell him: you are me now, you can't do away with me.

You do away with one another. All your genocides, your holocausts, your hatreds: what were they but reducing others to their genetic stuff? Kike-clone, nigger-clone, Spic and Chink and Mick. How much, he'd say, is actually disposable to you: how much, even of your selves, you throw carelessly away.

Sometimes it occurred to me that I was taking vengeance on existence through him. Its uncertainties, its impossibilities. Perhaps my real desire was to prove that all the quandaries of consciousness were valid, no matter how much self-inflicted. We had made with him the antithesis to quandary: that which comes to birth exactly from self-consciousness, the part of self that always stands aside beholding, registering the self, even in its quandary, even in its pain.

Don't worry, he would say, my dying won't affront me as yours will you; I won't feel something's *stolen* from me, as you do. Death will still remain a human treasure.

Is there anything I could impart to him other than despair?

You want perfection, he would tell me, you human creatures, but not too much of it, not enough of anything that possibly might change your nature. You so love your nature, the way you're so defined by your biologic past: you meant me to sublimate this for you.

Your ambition, he would say, so outstrips your moral force, yet your ambitions are what you cultivate. What am I, he would say, but one more instance of your rhetoric, of your pathetic illusion-making?

But the *spirit,* I might ask him: is the human spirit nothing more than all the false equations, false confusions, anxieties, misapprehensions, horrid histories, of which it seems to be composed? Is there nothing past all that? Are we merely monsters?

Always you conceived of *me,* he'd say, as monster: because I represent the illusion of perfectibility you believed should be in spirit, yet never is. And that illusion is an abnegation of what you think is worthiest in you. You'd like to think that what can't be stated in a formula about you is precisely what you are, yet you can't keep yourself from generating formulae.

But why really must you leave me?

You said yourself, he'd tell me, it's the wounds, because by definition I can't have your wounds, and so by definition must affront you, disappoint you. You define yourself by character, and character by aberrations from a norm. Don't I represent a norm? Aren't I meant to *be* a norm? To the degree that I develop an identity, I become more threatening to you: a generated self, a monster. I'm so circumscribed by all your notions of monstrosity that I can hardly move. And yet without your wounds, I have no reason to.

I'd thought, I'd say, that because of how we'd made you, with our minds, that you'd have access with *your* mind to mysteries that denied us. I had thought you'd stand out in the night and hear the chording of the stars which to us are silent; I'd thought you'd hear the octaves that unfold from cell to cosmos.

Oedipus and angel: how much you wished of me, a creature generated from such contaminated wants.

Perhaps we wanted you to rectify our foolishness, our weakness. Perhaps we thought that if the definition of the species changed, we would love each other more. Perhaps what we wished from you was *love.*

Self-veneration.

To complete ourselves in love by means of you.

Not an angel, but a minotaur, a monster.

I imagine rending sadness. That I ask him not to go, and yet he goes.

But what will be your legacy to us?

Self-love. Self-loathing.

And what of us will you take with you?

Loathing. Love.

Little C

Martha C. Nussbaum

He seized her by the arm, unable to speak. Trembling, she
tried to run after Jeannie and Jeannette, but he drew her
back as if by force, and made her return with him. And
Madeleine, seeing how his will gave him the daring to resist
hers, understood, far better than by any words, that it was
no longer her child, the foundling boy, it was François her
lover who was walking by her side.

GEORGE SAND, *François de Champi*

You weren't there any more. I don't know why, but I know you
were gone, just not there any more, and I was frozen with grief.
(You know how it would be. I was eating lots of salad, running
four miles a day, writing several articles—but still, my heart was a
heavy cold block.)

Then, one day, our friends at the institute told me that they had
a surprise for me. They had seen how unhappy I was, and they
wanted to bring me something. I heard my apartment doorbell
ring. I opened the door to the elevator. There, nestling in a purple
laundry basket, surrounded by rushes and daffodils, wrapped in a
green plaid pajama top, was the baby clone. He looked up at me
with the defiant smile of baby Hercules getting ready to throttle the

serpents. His hands were already big enough to hold a tennis racket, and his thighs showed signs of promise.

Little C.

I picked him up and embraced him, declaring that henceforth he would live with me in my house as my very own child.

How I loved Little C. I would hold him so hopefully in my arms, thinking that he would soon become you. When his eyes turned from baby blue to a deeper gray-blue with flecks of yellow, a wonderful joy began to seep into my heart. I never tired of nursing him (through medical advances this was possible). I felt the baby lips around my nipple, and I imagined, as the milk flowed out, how the new sensation would please you. How eagerly I watched his hands wave in the air, describing ever more articulate and commanding gestures.

I gave Little C many baths. His legs pounded the water with rebellious strength, and he laughed with defiance as he covered me with water.

Often, in the late afternoon, Little C lay beside me as I rested, and made his happy baby gurgling noises. "Very good, Little C," I said.

But I also teased him, saying, "When are you going to talk about your own ideas, about global redistribution, and the shortcomings of utilitarianism? Move along quickly, Little C, for there is something lacking in this relationship."

I examined the different parts of his body, so white and soft, growing bigger under my care and nutrition. And I teased him again, saying, "Very good, Little C, but where is the lovely large body of which I am especially fond? Move along more rapidly, Little C."

In the early evening we sat at the window overlooking the lake, and we watched the evening light grow paler, the flecks of gold changing to swirls of rose and gray. Little C lay contented in my arms, and I sang to him my favorite songs, such as "Caro mio ben," and "Rêve d'Amour." Knowing you well, I watched for signs of boredom; but Little C listened to the end, contentedly gurgling.

As time went on, Little C got bigger and more wonderful. He walked at ten months, and very soon he showed a quickness and poise beyond his years. His strong legs pounded the floor as he ran, and I could see the muscles in his thighs growing rapidly. I moved to a house with a large yard, so that Little C could run and jump. A natural athlete, I told my colleagues at the institute, who were not surprised. They smiled at the extravagance of my maternal praise.

As Little C played, I would watch his movements closely, to see whether he had begun to develop that sloping posture, right shoulder slightly lower than left, by which I could recognize you three miles off or from the air at 10,000 feet. The shoulder slopes a little as if its heavier muscles are pulling it down, and the back twists ever so slightly to the right, portending grim prospects for the opponent. I would think I saw its signs, although Little C had never been on a tennis court. (Indeed, at that time he showed a strong preference for youth soccer.) I loved the light in his multicolored eyes, the jutting defiance of his jaw, the deft and rapid movements of his feet.

When Little C was eight, I began to take him with me to the opera. It is only an experiment, I told myself, and I will stop it if he shows any signs of boredom. How happy I was that Little C reacted well. First *Hansel and Gretel,* and soon even *The Magic Flute,* although he expressed disapproval of its images of racial and sexual inequality. We spent blissful intermissions together discussing the two principles of global justice with reference to Monostatos and the Queen of the Night. (Little C did not use that philosophical language, of course, but I noted with pleasure that he seemed to gravitate naturally toward the core ideas.) Soon Little C was asking to be taken to the opera on a regular basis.

"See," I imagined myself saying to you, "Little C likes classical music a great deal, and opera most particularly. So it wasn't in the genes, was it? The principles of global justice were in the genes, but the anti-opera principle was not." And I couldn't wait for him to develop a taste for Verdi, and even perhaps Wagner. Life looked very promising at that time.

As mothers go, I tended to the Proustian. I would promise Little C a bedtime kiss, and when, like Marcel, he implored me to stay longer and read to him, I would come into his room and read for hours. Among our favorite books was George Sand's *François le Champi,* the book Marcel's mother reads to him when she stays up with him all night. Little C was entranced by the story of the young miller's wife who finds a foundling boy in the field and decides to bring it up as her own child. He especially loved the part where Madeleine, looking at the poor cold wretched abandoned boy, asks him what his name is. "They call me François the foundling." "François le Champi." At that name (so indicative, had he known it, of his own condition), Little C's eyes grew bright with joy, and he liked to repeat the name in French, Champi, as if it were his own. "Then, Little C," I continued, "Madeleine looked at the little Champi with a gaze full of compassion. She picked him up, and announced that henceforth he would live in her house as her very own child."
And that was the manner in which I revealed to Little C his strange origin. One day, "You, my love," I told him, "are that Champi. For I found you: not in a field, but at my door, lying in a purple laundry basket, surrounded by rushes and daffodils, wrapped in a green plaid pajama top. You had as excellent a smile as baby Hercules getting ready to throttle the two serpents, and your thighs already showed signs of their current strength. I picked you up in my arms and embraced you, announcing that henceforth you would live in my house as my very own child." After that, Little C never tired of hearing that story. He requested it almost every day.

At this time concern for propriety made me refrain from re-vealing to Little C the ending of Sand's narrative. How the miller's wife, abandoned by her husband, grows very close to the foundling boy. And how one day, after years of intimate domestic life, she no-tices that the Champi is a grown man, and amazing in beauty. How he shows adult defiance of her will, and seizes her in a passionate embrace. No, I concealed those portions of the book, and ended my readings with the Champi's boyhood. But after Little C went to sleep, I read to myself, frequently, the scene where Madeleine and François embrace, and she recognizes, with joy, feeling the power of a mature and independent will, that the child she has raised as her own will henceforth be her lover and her husband. While Little C slept, I would look out at the moon over the black lake, and think of the happiness in store.

One day when Little C was ten, he said to me, "Mother, green is such a beautiful color. Why do you never wear green dresses?" As-tonished, I replied, "What a ridiculous question. Because you hate green." But I was wrong, for Little C did not hate green. So I got out the green Armani suit that was hanging in my closet unused when you were there, and the green silk shell that looks so nice under several jackets, and I wore them for Little C's pleasure, and for my own. And I thought, with a softly sinking feeling in my stomach, "Why does Little C like green? Surely I look better in blue."

Then one day, shortly after this, I said, "Little C, please clean up your room." And, since I asked very gently, giving no incentives for defiance, Little C obeyed. And every day from then on, without my even having to ask, the room was clean. Papers in neat piles on the desk, books on the shelf, socks in the laundry, pajamas hung up on a hook behind the door, cups and plates neatly stacked in the dish-washer. I watched with approval and gentle encouragement. And the ice of grief began to grow again in my heart.

People from the institute, who knew the story of Little C, came to marvel at the room, as at a wonder. Some approved of the alteration. The institute's chief economist, fastidious, felt himself released from a long-standing disgust. Our director, too, was relieved and gratified. But others—your philosopher friends in the global justice project—began to sneak behind my back and say, "Little C, it's okay to leave your socks on the floor." "Little C, this half-empty can of Diet Coke looks really great turned upside down on your desk." "Little C, let's get out some papers and pile them up on the floor." But Little C said, "My mother asked me always to keep my room clean." And so he would refuse them. And they, too, began to grieve.

How, I thought, had I produced a child so pliable, so lacking in willfulness? Had I nursed him too often? Bathed him too tenderly? Sung too many soft French love songs? Where was my heroic child, fit to leap over all obstacles, including those imposed by his mother? Could it be that the secrets of making love to you were so well known to me, while the secrets of producing you were unknown completely?

At this time, my heart began to alter. Oh I was a good mother still, and I did the things that good mothers do. But the wild hopefulness and joy slowly drained out of our daily interactions. I did not sing or read as often to Little C, even though the knowledge of his individuality made it rational to sing all the more, since he seemed inclined to cultivate the musical talents that you spurned. In lieu of singing, I arranged for piano lessons, and Little C duly became a fine musician.

I no longer looked for the rightward slope in the shoulder. I noted that, in fact, Little C had a definite preference for soccer. He showed no inclination for either tennis or any other racket sport. For my part, I could not find much enthusiasm for soccer, a game I have always disliked. Little C's deft motions up and down the field began to bore me, since I saw no daring in them.

Or perhaps it was the body of Little C that failed to hold my interest, so skinny and light, with neither muscular shoulders nor thighs of any substance.

During this same period a change also came over Little C. His multicolored eyes grew more subdued, losing their flashes of yellow light. His humor, once so wild and extravagant, subsided, as if beneath a weight. His running, though indeed exceedingly deft, lost the edge of exuberance that made people speak of rare athletic gifts.

Instead, as he grew into a tall boy of twelve and then thirteen, he poured his emotions into the piano, practicing for hours, with a gloomy intensity that astonished those who had previously known him. From Bach fugues and Mozart sonatas, he moved on, seeking pensive solemn music, music of lost love and a world of unavailable ease and grace. Satie, Debussy, Ravel—a world of pale moonlight, where the joy is so distant that it exists only in fantasy. The house vibrated with the haunting notes of "La Cathédrale Engloutie," as the cathedral, buried under the ocean floor, rises gloriously into the light for one moment—and sinks again beneath the waves.

Occasionally, charmed by the music itself, I allowed myself to sing while Little C played. Songs of Duparc and Fauré, songs with lines such as "Exiled from a golden sky where your beauty flourishes." And "Beyond the roof, how blue, how calm, the sky is." I felt as I sang that I could see your face through the music, and at those moments I loved Little C for bringing you closer. Little C was happy then. Increasingly, he sought out the piano.

Through his music he won much acclaim. People spoke of a rare poetic sensibility in one so young.

A time came when Little C was seventeen, and due to leave home shortly, to continue his musical studies at Juilliard. For although he was a fine academic student, he cared deeply for nothing but music, and he could not be truly happy unless he was playing something delicate and sad. The night before his departure for New York, we went for a last celebratory evening, mother and son, at the opera.

By chance, they were performing *Don Carlo,* and we sat together in silence through the Fontainebleau scene. Elisabetta and Carlo, finding that they are fated to be mother and son rather than lovers, sang of the horrible pain of their renunciation. "L'hora fatale è suonata," the fateful hour has sounded, and love is doomed forever. Yes, I thought. Doomed to be mother and son, forever.

In the intermission, Little C stood beside me, and I smiled up at him. By now, he was six foot three, although he retained his skinny tense physique and, being an athlete no longer (since he feared injury to his hands), he still had no shoulders to speak of. His multicolored eyes gleamed with a quiet, no longer a heroic light. We analyzed the performance, as was our habit.

Then Little C looked at me with the grave sadness that had by now become his characteristic expression.

"Mother," he said to me, "I see that I do not make you happy."

"It is true that I am not happy, Little C," I said to him. "But it has nothing to do with you."

"I have always tried so hard to please you, Mother," he said. "But no matter what I do or say, you are always just a little sad, and your eyes look at me as if you are thinking of something else."

"That is true, Little C," I said. "It is not your fault, but it is the truth."

"What are you thinking about, Mother, when that sad lost expression comes into your eyes? I wish I could know, because perhaps then I would be able to make you happy."

"It is a long long story, Little C, and you cannot know it."

"And that baby in the laundry basket, surrounded with rushes and daffodils, wrapped in a green plaid pajama top. Am I that baby?"

"You are indeed that baby. My Champi. My Little C."

"Why, then, do you not love me the way Madeleine loved her grown François?"

"Because each story has its own ending, and no person is exactly like any other."

"Am I then less lovable than François was?"

"You are the best Little C the world has ever known. Now let us go and take our seats. The intermission is almost over, and the second act is very fine."

Contributors' Notes

Felicia Ackerman is professor of philosophy at Brown University, as well as a writer whose short stories, many of which deal with issues in medical ethics, have appeared in ten magazines and one O. Henry Awards collection.

Dan Brock is the Charles C. Tillinghast, Jr., University Professor; professor of philosophy and biomedical ethics; and director of the Center for Biomedical Ethics, at Brown University.

Richard Dawkins is the Charles Simonyi Professor of Public Understanding of Science at Oxford University. His latest book is *Climbing Mount Improbable* (1997), published by W. W. Norton.

Wendy Doniger is the Mircea Eliade Professor of the History of Religion at the University of Chicago. She is the author of *Other Peoples' Myths, The Implied Spider,* and several Penguin translations from the Sanskrit.

Andrea Dworkin is the author of *Pornography: Men Possessing Women, Intercourse, Life and Death,* and the novels *Mercy* and *Ice and Fire.*

Jean Bethke Elshtain is the Laura Spelman Rockefeller Professor of Social and Political Ethics at the University of Chicago. She is the author, most recently, of *Democracy on Trial* (1995) and *Augustine and the Limits of Politics* (1996). She is also a contributing editor to the *New Republic.*

Richard A. Epstein is the James Parker Hall Distinguished Service Professor of Law at the University of Chicago.

William N. Eskridge, Jr., is professor of law at Georgetown University Law School and the author of *The Case for Same-Sex Marriage, Dynamic Statutory Interpretation,* and *Cases and Materials in Legislation: Statutes and the Creation of Public Policy* (with Philip Frickey).

Stephen Jay Gould is the author of sixteen books, including, for Norton, such international bestsellers as *Ever Since Darwin, The Panda's Thumb, Bully for Brontosaurus, Wonderful Life,* and *Eight Little Piggies,* and most recently, for Random House, *Questioning the Millennium.* Winner of the American Book Award for Science and of the National Book Critics Circle Award for *The Mismeasure of Man,* he teaches geology, biology, and the history of science at Harvard University.

George Johnson is a science correspondent for the *New York Times* writing from Santa Fe, New Mexico. His books include *Fire in the Mind: Science, Faith, and the Search for Order, In the Palaces of Memory: How We Build the Worlds Inside Our Heads, Machinery of the Mind: Inside the New Science of Artificial Intelligence,* and *Architects of Fear: Conspiracy Theories and Paranoia in American Politics.*

William Ian Miller is a professor of law at the University of Michigan. He is the author of *The Anatomy of Disgust* (1997), *Humiliation* (1993), and *Bloodtaking and Peacemaking* (1990).

The National Bioethics Advisory Commission (NBAC) was created by executive order of President Clinton in October 1995 to offer recommendations to the National Science and Technology Council, as well as to any government agencies concerned with human biological and behavioral research, on the ethical treatment of human research subjects. The NBAC was also charged with considering the issues that human genetics research raised and any other significant bioethical issues they or the Congress or the public identified. The order encouraged the Commission to hold hearings, develop reports, commission papers, and form subcommittees as needed. The Commission's members are: Harold T. Shapiro, President of Princeton University; Patricia Backlar, Research Associate Professor for Bioethics at Portland State University; Arturo Brito, Assistant Professor of Clinical Pediatrics at University of Miami School of

Medicine; Alexander M. Capron, Henry W. Bruce Professor of Law and
University Professor of Law and Medicine at University of Southern Cal-
ifornia Law Center; Eric J. Cassell, Professor of Public Health at Cornell
Medical College; R. Alta Charo, Associate Professor of Law and Medical
Ethics at University of Wisconsin Schools of Law and Medicine; James F.
Childress, Kyle Professor of Religious Studies and Professor of Medical
Education at University of Virginia; David R. Cox, Professor of Genet-
ics and Pediatrics at Stanford University School of Medicine; Rhetaugh G.
Dumas, Vice Provost Emerita and Dean Emerita and Lucille Cole Pro-
fessor of Nursing at University of Michigan; Laurie M. Flynn, Executive
Director of National Alliance for the Mentally Ill; Carol W. Greider, As-
sociate Professor of Molecular Biology and Genetics, Johns Hopkins Uni-
versity School of Medicine; Steven H. Holtzman, Chief Business Officer,
Millennium Pharmaceuticals; Bette O. Kramer, Founding President,
Richmond Bioethics Consortium; Bernard Lo, Director, Program in
Medical Ethics, University of California, San Francisco; Lawrence H.
Miike, Director, State Department of Health, Hawaii; Thomas H. Mur-
ray, Professor and Director, Center for Biomedical Ethics, Case Western
Reserve School of Medicine; and Diane Scott-Jones, Professor of Psy-
chology, Temple University.

Martha C. Nussbaum is Ernst Freund Professor of Law and Ethics at the
University of Chicago, with appointments in the Law School, the Phi-
losophy Department, and the Divinity School. Among her books are *The
Fragility of Goodness, Love's Knowledge,* and *For Love of Country* (ed.).

Adam Phillips is principal child psychotherapist in the Wolverton Gar-
dens Child and Family Consultation Center in London and a member of
the Guild of Psychotherapists. He is the author of *On Kissing, Tickling, and
Being Bored* and *On Flirtation,* and has edited such volumes as *Charles Lamb:
Selected Prose, Richard Howard: Selected Poems,* and *A Philosophical Enquiry* by
Edmund Burke.

Eric A. Posner is professor of law at the University of Chicago.

Richard A. Posner is chief judge of the United States Court of Appeals for
the Seventh Circuit, senior lecturer at the University of Chicago Law
School, and the author of *Sex and Reason.*

Barbara Katz Rothman is professor of sociology at the City University of New York. Her books include *The Tentative Pregnancy,* and the forthcoming *Of Maps and Imaginations: Confronting the Human Genome.*

Edward Stein is the author of *Without Good Reason: The Rationality Debate in Philosophy and Cognitive Science* (Oxford, 1996) and the editor of *Forms of Desire: Sexual Orientation and the Social Constructionist Controversy.* His book *Sexual Desires: Science, Theory, and Ethics* is forthcoming in 1998 from Oxford. He has also written many articles on philosophy and on lesbian and gay studies.

Cass R. Sunstein is the Karl N. Llewellyn Distinguished Service Professor of Jurisprudence in the Law School and Department of Political Science at the University of Chicago. He is the author of *Free Markets and Social Justice* (1997), *Legal Reasoning and Political Conflict* (1996), and *Democracy and the Problem of Free Speech* (1993).

David Tracy is Distinguished Service Professor at the University of Chicago and the Andrew T. and Grace McNichols Professor of Catholic Studies at the Divinity School of the University. He is also a member of the Committee on Social Thought. His books include *Plurality and Ambiguity: Hermenueutics, Religion, Hope.*

Laurence Tribe is the Tyler Professor of Constitutional Law at Harvard University, a fellow of the American Academy of Arts and Sciences, holder of numerous honorary degrees, author of many books and articles, and a leading advocate before the Supreme Court, where he has argued and won many landmark cases.

Lisa Tuttle, an American living in Scotland, is the author of many short stories in the field of science fiction, fantasy, and horror. She has also written several novels, including *Windhaven* (co-authored with George R. R. Martin) and, most recently, *The Pillow Friend,* as well as books for children and the nonfiction *Encyclopedia of Feminism.*

C. K. Williams's most recent books of poetry are *Selected Poems* and *The Vigil.* He teaches at Princeton University.

Ian Wilmut was awarded the Ph.D. from the University of Cambridge for research on the Deep Freeze Preservation of Boar Semen. Subsequent research in Cambridge led to the birth of the first calf from a frozen embryo ("Frosty" in 1973). His work at the Animal Breeding Research Organization has been concerned with developing techniques of multiple ovulation/embryo transfer in sheep and cattle, and he was a joint leader of the team that produced transgenic sheep at Roslin. Over the past five years, his research has been focused on the factors regulating embryo development in sheep after nuclear transfer, work leading to the first birth of live lambs from embryo-derived cells and then to the birth of lambs derived from fetal and adult cells, most famously, Dolly.